CMI Lecture Series in Mathematics 1

Lectures on the structure of algebraic groups and geometric applications

CMI Lecture Series in Mathematics

Editors

Vikraman Balaji, Chennai Mathematical Institute, Chennai.
Rajeeva Karandikar, Chennai Mathematical Institute, Chennai.
Pramathanath Sastry, Chennai Mathematical Institute, Chennai.

Volumes published so far

1. Lectures on the structure of algebraic groups and geometric applications, M. BRION, PREENA SAMUEL, V. UMA

Lectures on the structure of algebraic groups and geometric applications

Michel Brion
Institut Fourier, Grenoble, France

Preena Samuel
Chennai Mathematical Institute, Chennai, India.

V. Uma
Indian Institute of Technology Madras, Chennai, India.

Published by
Hindustan Book Agency (India)
P 19 Green Park Extension
New Delhi 110 016
India

email: info@hindbook.com
http://www.hindbook.com

ISBN 978-93-80250-46-5

Preface

The theory of algebraic groups has chiefly been developed along two distinct directions: linear (or, equivalently, affine) algebraic groups, and abelian varieties (complete, connected algebraic groups). This is made possible by a fundamental theorem of Chevalley: any connected algebraic group over an algebraically closed field is an extension of an abelian variety by a connected linear algebraic group, and these are unique.

In these notes, we first expose the above theorem and related structure results about connected algebraic groups that are neither affine nor complete. The class of anti-affine algebraic groups (those having only constant global regular functions) features prominently in these developments. We then present applications to some questions of algebraic geometry: the classification of complete homogeneous varieties, and the structure of homogeneous (or translation-invariant) vector bundles and principal bundles over abelian varieties.

While the structure theorems presented at the beginning of these notes go back to the work of Barsotti, Chevalley and Rosenlicht in the 1950's, all the other results are quite recent; they are mainly due to Sancho de Salas [Sal03, SS09] and the first-named author [Bri09, Bri10a, Bri11, Bri12]. We hope that the present exposition will stimulate further interest in this domain. In Chapter 1, the reader will find a detailed overview of the contents of the subsequent chapters as well as some open questions.

These notes originate in a series of lectures given at Chennai Mathematical Institute in January 2011 by the first-named author. He warmly thanks all the attendants of the lectures, especially V. Balaji, D. S. Nagaraj and C. S. Seshadri, for stimulating questions and comments; the hospitality of the Institute of Mathematical Sciences, Chennai, is also gratefully acknowledged. The three authors wish to thank Balaji for having prompted them to write up notes of the lectures, and encouraged them along the way; thanks are also due to T. Szamuely for his very helpful comments and suggestions on a preliminary version of these notes.

Contents

Chapter 1

Overview

Notation and conventions. Throughout these notes, we consider algebraic varieties and schemes over a fixed algebraically closed base field k. Unless otherwise stated, schemes are assumed to be of finite type over k, and points are assumed to be closed, or equivalently k-rational. For a scheme X, we denote by $\mathcal{O}(X)$ the algebra of global sections of the structure sheaf $\mathcal{O}_X$. A *variety* is a separated integral scheme.

A *group scheme* G is a scheme equipped with morphisms $m : G \times G \to G$ (the multiplication), $i : G \to G$ (the inverse) and with a point e_G (the neutral element) which satisfy the axioms of a group. The *neutral component* of G is the connected component containing e_G, denoted as G^o; this is a normal subgroup scheme of G, and the quotient G/G^o is a finite group scheme.

An *algebraic group* is a group scheme which is smooth, or equivalently, reduced; by a subgroup (scheme) of G, we always mean a closed subgroup (scheme). Any group scheme G contains a largest algebraic subgroup, namely, the underlying reduced subscheme G_{red}.

An *abelian variety* A is a complete connected algebraic group. It is well-known that such an A is a projective variety, and its group law is commutative; we denote that law additively, and the neutral element as 0_A. For any non-zero integer n, we denote as n_A the multiplication by n in A, and as A_n its scheme-theoretic kernel; recall that n_A is an isogeny of A, *i.e.*, a finite surjective homomorphism.

As standard references, we rely on the books [Har77] for algebraic geometry, [Spr09] for linear algebraic groups, and [Mum08] for abelian varieties; for the latter, we also use the survey article [Mil86]. We refer to [DG70] for group schemes.

1.1 Chevalley's structure theorem

The following theorem was first stated by Chevalley in 1953. It was proved in 1955 by Barsotti [Bar55] and in 1956 by Rosenlicht [Ros56]; both used the language and methods of birational geometry à la Weil.

Theorem 1.1.1 *Let G be a connected algebraic group. Then G has a largest connected affine normal subgroup G_{aff}. Further, the quotient group G/G_{aff} is an abelian variety.*

We shall present an updated version of Rosenlicht's proof of the above theorem in Chapter 2. That proof, and some further developments, have also been rewritten in terms of modern algebraic geometry by Ngô and Polo [NP11], during the same period where this book was completed.

In 1960, Chevalley himself gave a proof of his theorem, based on ideas from the theory of Picard varieties. That proof was later rewritten in the language of schemes by Conrad [Con02].

Chevalley's theorem yields an exact sequence

$$1 \to G_{\mathrm{aff}} \to G \xrightarrow{\alpha} A \to 1, \tag{1.1}$$

where A is an abelian variety; both G_{aff} and A are uniquely determined by G. In fact, α is the *Albanese morphism* of G, *i.e.*, the universal morphism from G to an abelian variety, normalized so that $\alpha(e_G) = 0_A$. In particular, A is the Albanese variety of G, and hence depends only on the variety G.

In general, the exact sequence (1.1) does not split, as shown by the following examples. However, there exists a smallest lift of A in G, as will be seen in the next section.

Example 1.1.2 Let A be an abelian variety, $p : L \to A$ a line bundle, and $\pi : G \to A$ the associated principal bundle under the multiplicative group $\mathbb{G}_m$ (so that G is the complement of the zero section in L). Then G has a structure of an algebraic group such that π is a homomorphism with kernel $\mathbb{G}_m$, if and only if L is algebraically trivial (see *e.g.* [Ser88, VII.16, Theorem 6]). Under that assumption, G is commutative, and its group structure is uniquely determined by the choice of the neutral element in the fibre of L at 0_A. In particular, the resulting extension

$$0 \to \mathbb{G}_m \to G \to A \to 0$$

is trivial if and only if so is the line bundle L. Recall that the algebraically trivial line bundles on A are classified by $\mathrm{Pic}^0(A) =: \hat{A}$, the *dual abelian variety*. Thus, $\hat{A}$ also classifies the extensions of A by $\mathbb{G}_m$.

Next, let $q : H \to A$ be a principal bundle under the additive group $\mathbb{G}_a$. Then H always has a structure of an algebraic group such that q is a homomorphism; the group structure is again commutative, and uniquely determined by the choice of a neutral element in the fibre of q at 0_A. This yields extensions

$$0 \to \mathbb{G}_a \to H \to A \to 0$$

classified by $\mathrm{H}^1(A, \mathcal{O}_A)$; the latter is a k-vector space of the same dimension as that of A (see [Ser88, VII.17, Theorem 7] for these results).

Remark 1.1.3

(i) Chevalley's Theorem still holds over any perfect field, by a standard argument of Galois descent. Also, any connected group scheme G over an arbitrary field has a smallest connected affine normal subgroup scheme H such that G/H is an abelian variety (see [BLR90, Theorem 9.2.1]; its proof proceeds by reduction to the case of a perfect field). In that statement, the smoothness of G does not imply that of H; in fact, Chevalley's theorem fails over any non-perfect field as shown by an example of Raynaud, see [SGA3, Exposé XVII,

Corollaire C.5]. The preprint [Tot11] contains further developments on the structure of algebraic groups over non-perfect fields.

(ii) It would be very interesting to extend Chevalley's theorem to the setting of a smooth, connected group scheme over a smooth base of dimension 1, *e.g.*, the spectrum of the power series ring $k[[t]]$. Note that a direct generalization of that theorem is incorrect, as there exist such group schemes with generic fibre an abelian variety and special fibre a torus.

1.2   A dual structure theorem and the Rosenlicht decomposition

The following dual statement to Chevalley's theorem is due to Rosenlicht, see [Ros56, Corollaries 3, p. 431 and 5, p. 440]). The (rather easy) proof will be presented in Chapter 3.

Theorem 1.2.1 *Let G be an algebraic group. Then there exists a smallest normal subgroup scheme H of G such that the quotient G/H is affine. Moreover, $\mathcal{O}(H) = k$, and H is the largest subgroup scheme of G satisfying that property; it is in fact a connected algebraic group, contained in the center of G^o. Also, $\mathcal{O}(G/H) = \mathcal{O}(G)$; in particular, the algebra $\mathcal{O}(G)$ is finitely generated.*

We say that a scheme Z is *anti-affine* if $\mathcal{O}(Z) = k$. It is easy to see that the above subgroup H is the largest anti-affine subgroup of G; we denote it as G_{ant}. Moreover, the quotient homomorphism $G \to G/G_{\mathrm{ant}}$ is the canonical morphism $G \to \operatorname{Spec} \mathcal{O}(G)$, called the *affinization morphism*. In particular, G_{ant} depends only on the variety G.

Remark 1.2.2

(i) The above theorem holds unchanged in the setting of group schemes (of finite type) over an arbitrary field, see [DG70, §III.3.8].

(ii) Like for Chevalley's theorem, it would be very interesting to extend the above theorem to the setting of group schemes over a smooth base of dimension 1. By [SGA3, Exposé VI B, Proposition 12.10], the affinization morphism still satisfies nice finiteness properties in this setting. But the fibres of this morphism are not necessarily anti-affine, as shown again by degenerations of abelian varieties to tori.

Example 1.2.3 As in Example 1.1.2, let $0 \to \mathbb{G}_m \to G \xrightarrow{\alpha} A \to 0$ be the extension associated to an algebraically trivial line bundle L. Then G is anti-affine if and only if L has infinite order in $\hat{A}$. (This follows from the facts that $\alpha_*(\mathcal{O}_G) \cong \oplus_{n \in \mathbb{Z}} L^n$, and that for any $M \in \hat{A}$, $\mathrm{H}^0(A, M) \neq 0$ if and only if M is trivial. See Section 5.3 for details and further developments).

Next, let $0 \to \mathbb{G}_a \to H \to A \to 0$ be the extension associated to a principal $\mathbb{G}_a$-bundle over A. When $\mathrm{char}(k) = 0$, this extension is anti-affine if and only if it is non-trivial, as we shall show in Section 5.4. In contrast, this extension is never anti-affine when $\mathrm{char}(k) = p \neq 0$. Indeed, multiplication by p on $\mathbb{G}_a$ is trivial so $\mathrm{Ext}^1(A, \mathbb{G}_a)$ is killed by p, giving us the following pull-back diagram (by the bilinearity of $\mathrm{Ext}^1(A, \mathbb{G}_a)$):

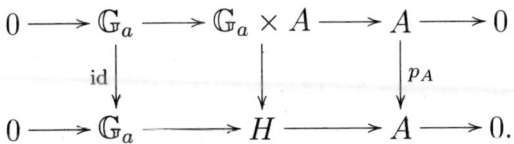

This readily gives $\mathcal{O}(H) \cong \mathcal{O}(\mathbb{G}_a \times A)^\Gamma \cong \mathcal{O}(\mathbb{G}_a)^\Gamma$, where Γ is a subgroup scheme of $\mathbb{G}_a \times A$, isomorphic to A_p (the kernel of p_A) via the second projection. Then Γ is finite; thus, there are non-constant regular functions on H.

We shall present a complete classification of anti-affine groups in Section 1.4.

The above theorems of Chevalley and Rosenlicht may be combined to give a decomposition of any connected algebraic group, also due to Rosenlicht (see [Ros56, pp. 440–441]):

Theorem 1.2.4 *Let G be a connected algebraic group, G_{aff} its largest connected affine normal subgroup, and G_{ant} its largest anti-affine subgroup. Then $G = G_{\mathrm{aff}}G_{\mathrm{ant}}$, and G_{ant} is the smallest subgroup scheme H of G such that $G = G_{\mathrm{aff}}H$, i.e., the restriction to H of the quotient homomorphism $\alpha : G \to G/G_{\mathrm{aff}}$ is surjective. Moreover, the scheme-theoretic intersection $G_{\mathrm{aff}} \cap G_{\mathrm{ant}}$ contains $(G_{\mathrm{ant}})_{\mathrm{aff}}$ as a normal subgroup with finite quotient.*

This result will be deduced in Chapter 3 from Theorems 1.1.1 and 1.2.1; we shall refer to the above decomposition as the *Rosenlicht decomposition*.

1.3 Structure of complete homogeneous varieties

Theorem 1.3.1 *Let X be a complete variety which is homogeneous under the action of some algebraic group. Then $X \cong Y \times A$, where A is an abelian variety and Y is a complete homogeneous variety under the action of an affine algebraic group. Moreover, the projections $X \to Y$, $X \to A$ are unique, and $\mathrm{Aut}^\circ(X) \cong \mathrm{Aut}^\circ(Y) \times A$ where A acts by translations on itself, and $\mathrm{Aut}^\circ(X)$ denotes the neutral component of the automorphism group scheme $\mathrm{Aut}(X)$.*

This result was first proved by Borel and Remmert in the setting of compact homogeneous Kähler manifolds (see [BR62]). The result in the generality discussed above is due to Sancho de Salas ([Sal03, Theorem 5.2]), who gave an explicit construction of the projections. In Chapter 4 we shall present another proof, based on the structure of algebraic groups.

In the above statement, one easily deduces from Borel's fixed point theorem that $Y \cong G/H$ for some semi-simple group G of adjoint type and some subgroup scheme H of G. In characteristic 0, such an H is just a parabolic subgroup of G; in other words, Y is a *flag variety*. In positive characteristics, however, there are many more such H (non-reduced versions of parabolic subgroups); the corresponding homogeneous spaces G/H are called *varieties of unseparated flags*. Here is the simplest example of such a variety:

Example 1.3.2 Let $Y \subset \mathbb{P}^2 \times \mathbb{P}^2$ (two copies of projective plane) be the hypersurface given by the equation $x^p x' + y^p y' + z^p z' = 0$, where x, y, z (resp. x', y', z') denote the homogeneous co-ordinates on the first (resp. second) $\mathbb{P}^2$. We have an action of the group $G := \mathrm{GL}_3$ on $\mathbb{P}^2 \times \mathbb{P}^2$ via the usual action on the first $\mathbb{P}^2$ and via the action

$$(a_{ij}) \cdot [v] := [(a_{ji}^p)^{-1} \cdot v]$$

on the second $\mathbb{P}^2$. This action preserves Y, and one may check that Y is homogeneous under G. The isotropy subgroup scheme H of the point $([1 : 0 : 0], [0 : 0 : 1]) \in Y$ can be written as follows:

$$\left\{ \begin{pmatrix} * & * & * \\ 0 & * & * \\ 0 & x & * \end{pmatrix} \; : \; x^p = 0 \right\}.$$

In particular, H is non-reduced, and the associated reduced scheme is a Borel subgroup B of G.

Define a morphism $\pi : Y \longrightarrow \mathbb{P}^2 \times \mathbb{P}^2$ by

$$([x, y, z], [x', y', z']) \longmapsto ([x^p, y^p, z^p], [x', y', z']).$$

Then the equation defining Y is sent to the incidence relation which defines the flag variety, G/B, in $\mathbb{P}^2 \times \mathbb{P}^2$. So π is a purely inseparable covering of this flag variety.

Also, Y is a hypersurface of bi-degree $(p, 1)$ in $\mathbb{P}^2 \times \mathbb{P}^2$, and hence the canonical bundle of Y is $\mathcal{O}_Y(p - 3, -2)$ (by the usual formula for canonical bundle of hypersurfaces). If $p \geq 3$ then the anti-canonical bundle is not ample, *i.e.*, Y is not Fano. Hence Y is not isomorphic to any flag variety G'/P'.

Remark 1.3.3 In the above example, we have $H = P_1 \cap G_{(1)} P_2$ where P_1 and P_2 denote the standard maximal parabolic subgroups of G, and $G_{(1)}$ stands for the kernel of the Frobenius endomorphism F of G. So this construction yields similar examples for all simple groups G of rank ≥ 2.

In fact, for $p > 3$, any subgroup scheme H of G such that G/H is complete is of the form $P_1 G_{(n_1)} \cap P_2 G_{(n_2)} \cap \cdots$ where $P_1, P_2, \ldots$ are pairwise distinct maximal parabolic subgroups of G and n_1,

n_2, ... are non-negative integers; here $G_{(n)}$ denotes the kernel of the iterated Frobenius morphism F^n (see [Wen93, HL93] for this result and further developments).

One may check that the family of all smooth hypersurfaces of bi-degree $(p, 1)$ in $\mathbb{P}^2 \times \mathbb{P}^2$ yields a non-trivial deformation of Y. In contrast, such deformations do not exist in the setting of flag varieties; see [Dem77, Proposition 5].

Also, we shall see in Proposition 4.3.4 that some varieties of un-separated flags have a non-reduced automorphism group scheme; this contrasts again with flag varieties, whose automorphism group scheme is a semi-simple algebraic group of adjoint type by the main result of [Dem77]. It would be of interest to describe the automorphisms and deformations of all varieties of unseparated flags.

1.4   Structure of anti-affine algebraic groups

Let G be an anti-affine algebraic group; then G is connected and commutative in view of Theorem 1.2.1. By Chevalley's theorem, G is an extension (1.1) of an abelian variety A by a connected commutative affine algebraic group G_{aff}.

By the structure of commutative affine algebraic groups (see [Spr09, Theorem 3.3.1]), we know that $G_{\text{aff}} \cong T \times U$, where T is a torus and U is a connected commutative unipotent algebraic group; moreover, T and U are unique. Thus, we get two exact sequences from the above exact sequence:

$$0 \to T \to G_s := G/U \to A \to 0, \tag{1.2}$$

$$0 \to U \to G_u := G/T \to A \to 0. \tag{1.3}$$

Further, one easily checks that

$$G \cong G_s \times_A G_u.$$

The group G_s in the first sequence (1.2) is an extension of an abelian variety by a torus; such a group is called a *semi-abelian variety*, and the extension is classified by a homomorphism of abstract groups $c: \hat{T} \to \hat{A}(k)$ given by $\chi \mapsto G_\chi$, where $\hat{T}$ denotes the group of characters of T, and $\hat{A}(k)$ denotes the group of k-valued points of the dual abelian variety; G_χ stands for the push-out of (1.2) via the character χ, given by the following commuting diagram

$$
\begin{array}{ccccccccc}
0 & \longrightarrow & T & \longrightarrow & G_s & \longrightarrow & A & \longrightarrow & 0 \\
& & \downarrow{\chi} & & \downarrow & & \downarrow{\mathrm{id}} & & \\
0 & \longrightarrow & \mathbb{G}_m & \longrightarrow & G_\chi & \longrightarrow & A & \longrightarrow & 0.
\end{array}
$$

In characteristic 0, the connected commutative unipotent group U is a *vector group*, i.e., a finite dimensional k-vector space regarded as an additive group. For an abelian variety A, there exists a universal extension of A by a vector group (see [Ros58, Proposition 11], and also §5.4):

$$0 \to H^1(A, \mathcal{O}_A)^* \to E(A) \to A \to 0. \qquad (1.4)$$

Thus, the extension (1.3) is obtained as the push-out of the universal extension (1.4) via a unique linear map $\gamma: H^1(A, \mathcal{O}_A)^* \to U$:

$$
\begin{array}{ccccccccc}
0 & \longrightarrow & H^1(A, \mathcal{O}_A)^* & \longrightarrow & E(A) & \longrightarrow & A & \longrightarrow & 0 \\
& & \downarrow{\gamma} & & \downarrow & & \downarrow{\mathrm{id}} & & \\
0 & \longrightarrow & U & \longrightarrow & G_u & \longrightarrow & A & \longrightarrow & 0.
\end{array}
$$

Theorem 1.4.1 *With notations as above, G is anti-affine if and only if*

(i) *when* $\mathrm{char}(k) > 0$, *c is injective and U is trivial;*

(ii) *when* $\mathrm{char}(k) = 0$, *c is injective and γ is surjective.*

This result is proved in Chapter 5. When k is the algebraic closure of a finite field, it implies easily that every anti-affine group is an abelian variety. Together with the Rosenlicht decomposition (Theorem 1.2.4), it follows that every connected algebraic group G is of the form $(G_{\mathrm{aff}} \times A)/\Gamma$, where A is an abelian variety and Γ is a finite group scheme of $Z(G_{\mathrm{aff}}) \times A$. The latter result is due to Arima [Ari60, Theorem 1], see also Rosenlicht [Ros61, Theorem 4].

1.5   Homogeneous vector bundles over abelian varieties

Let A be an abelian variety, and $p : E \to A$ a vector bundle. Then p is said to be *homogeneous* (or *translation-invariant*) if, for all $a \in A$, $\tau_a^*(E) \cong E$ as vector bundles over A, where $\tau_a : A \to A$ denotes the translation by a.

For example, the homogeneous line bundles are exactly those in $\hat{A}$ (see [Mil86, Proposition 10.1]).

The structure of homogeneous vector bundles is described by the following result, due to Matsushima [Mat59] and Morimoto [Mor59] in the setting of complex vector bundles over complex tori, and to Miyanishi [Miy73] and Mukai [Muk78, Theorem 4.17] in the present setting.

Theorem 1.5.1 *Let E be a vector bundle over A. Then the following are equivalent:*

 (i) E *is homogeneous.*

 (ii) E *is an iterated extension of algebraically trivial line bundles.*

 (iii) $E \cong \oplus_i L_i \otimes E_i$ *where L_i are pairwise non-isomorphic, algebraically trivial line bundles, and E_i are unipotent vector bundles.*

Here a vector bundle of rank n is said to be *unipotent* if it is an iterated extension of trivial line bundles; equivalently, the associated principal GL_n-bundle has a reduction of structure group to the maximal unipotent subgroup

$$\begin{pmatrix} 1 & * & * & \cdots & * \\ 0 & 1 & * & \cdots & * \\ 0 & 0 & 1 & \cdots & * \\ & & & \ddots & \\ 0 & 0 & 0 & \cdots & 1 \end{pmatrix}.$$

In Chapter 6, we present a proof of the above theorem based on algebraic group methods, also showing that unipotent vector bundles over A correspond to representations $\rho : \mathrm{H}^1(A, \mathcal{O}_A)^* \to$ GL_n (up to conjugation), via the associated bundle construction $E(A) \times^{\mathrm{H}^1(A, \mathcal{O}_A)^*} k^n$. On the other hand, the Fourier-Mukai correspondence yields an equivalence of categories between unipotent vector bundles and coherent sheaves on $\hat{A}$ with support at 0_A (see [Muk81, Theorem 2.2]); the above construction yields another approach to that result, presented in Section 6.5.

Remark 1.5.2 (i) If A is an elliptic curve then $\mathrm{H}^1(A, \mathcal{O}_A) \cong k$. In this case, the unipotent vector bundles of rank n (up to isomorphism) correspond to the nilpotent $n \times n$ matrices (up to conjugation); the latter are classified by their Jordan canonical form. Thus, the indecomposable unipotent vector bundles correspond to the nilpotent Jordan blocks of the form

$$\begin{pmatrix} 0 & 1 & 0 & \cdots & 0 \\ 0 & 0 & 1 & \cdots & 0 \\ & & & \ddots & \\ 0 & 0 & 0 & \cdots & 1 \\ 0 & 0 & 0 & \cdots & 0 \end{pmatrix}.$$

In particular, such a bundle U_n is uniquely determined by its rank n. This result is due to Atiyah [Ati57b] who also showed that every indecomposable (not necessarily homogeneous) vector bundle over A, of degree d and rank r, is the tensor product of the indecomposable vector bundle U_n and of a simple vector bundle $S_{d',r'}$ of degree d' and rank r', where $n := \mathrm{g.c.d}(d, r)$, $d' := d/n$ and $r' := r/n$; moreover, $S_{d',r'}$ is uniquely determined by (d', r') up to tensoring with a line bundle of degree 0.

(ii) For an arbitrary abelian variety A of dimension g, classifying the unipotent vector bundles of rank n reduces similarly to classifying the n-dimensional representations of the vector group $\mathrm{H}^1(A, \mathcal{O}_A)^*$; the latter has dimension g as well. Choosing a basis of $\mathrm{H}^1(A, \mathcal{O}_A)^*$, this amounts to classifying the g-tuples of nilpotent $n \times n$ matrices up to simultaneous conjugation. The latter classification is a long-standing open question of representation theory, already for $g = 2$.

(iii) Given a connected algebraic group G with Albanese variety A, and a finite-dimensional representation $\rho : G_{\mathrm{aff}} \to \mathrm{GL}(V)$, the associated vector bundle $E_V := G \times^{G_{\mathrm{aff}}} V \to G/G_{\mathrm{aff}} = A$ is clearly homogeneous. In fact, the assignment $(\rho, V) \mapsto E_V$ yields a functor from the category of representations of G_{aff} to the category of homogeneous vector bundles on A, which preserves exact sequences, tensor products and duals. Can one reconstruct G from this functor, by generalizing the classical Tannaka duality for affine algebraic groups?

1.6 Homogeneous principal bundles over an abelian variety

Given an algebraic group G and an abelian variety A, a principal G-bundle $\pi : X \to A$ is said to be *homogeneous* if $\tau_a^*(X) \cong X$ as principal bundles over A, for all $a \in A$ (see Section 6.1 for details on principal bundles). In Chapter 7, we obtain a classification of these bundles by adapting the approach of the previous chapter:

Theorem 1.6.1 *Let G be a connected affine algebraic group. Then there is a bijective correspondence between homogeneous G-bundles $\pi : X \to A$ and pairs consisting of the following data:*

(i) *an exact sequence of commutative group schemes $0 \to H \to \mathcal{G} \to A \to 0$, where H is affine and $\mathcal{G}$ is anti-affine,*

(ii) *a faithful homomorphism $\rho : H \to G$, uniquely determined up to conjugacy in G.*

This correspondence assigns to any pair as above, the associated bundle $\pi : G \times^H \mathcal{G} \to \mathcal{G}/H = A$, where H acts on $\mathcal{G}$ by multiplication, and on G via right multiplication through ρ.

Moreover, the group scheme of bundle automorphisms, $\mathrm{Aut}_A^G(X)$, is isomorphic to the centralizer $C_G(H)$ of the image of H in G.

Remark 1.6.2 An exact sequence $0 \to H \to \mathcal{G} \to A \to 0$ as in (i) above (*i.e.*, H is affine and $\mathcal{G}$ is anti-affine) is said to be an *anti-affine extension*. These can be easily classified by adapting the arguments of the classification of anti-affine groups; the result is stated in Theorem 5.5.3.

However, classifying the data of (ii), *i.e.*, the conjugacy classes of faithful homomorphisms $\rho : H \to G$ where H is a prescribed commutative group scheme and G a prescribed connected affine algebraic group, is an open question (already when $H = \mathbb{G}_a^2$ and $G = \mathrm{GL}_n$, as we saw in Remark 1.5.2).

When $G = \mathrm{GL}_n$, one easily shows that a G-bundle $\pi : X \to A$ is homogeneous if and only if so is the associated vector bundle $X \times^G k^n \to A$. Then Theorem 1.6.1 gives back the structure of homogeneous vector bundles; for example, the decomposition in Theorem 1.5.1(iii) follows from the fact that any representation $\rho : H \to \mathrm{GL}_n$ of a commutative group scheme is the direct sum of its generalized eigenspaces.

For an arbitrary group G, we obtain a characterization of homogeneity in terms of associated vector bundles:

Theorem 1.6.3 *Let G be a connected affine algebraic group. Then the following conditions are equivalent for a G-bundle $\pi : X \to A$:*

(i) π is homogeneous.

(ii) For any representation $\rho : G \to \mathrm{GL}(V)$, the associated vector bundle $p : E_V = X \times^G V \to A$ is homogeneous.

(iii) For any irreducible representation $\rho : G \to \mathrm{GL}(V)$, the associated vector bundle is homogeneous.

(iv) For some faithful representation $\rho : G \to \mathrm{GL}(V)$ such that the variety $\mathrm{GL}(V)/\rho(G)$ is quasi-affine, the associated vector bundle is homogeneous.

A representation satisfying condition (iv) exists always (see Lemma 7.2.3). If G is reductive, then this condition holds for any faithful representation; the resulting characterization of homogeneous principal bundles is analogous to that obtained by Biswas and Trautmann ([BT10, Theorem 1.1]) in the setting of G-bundles over flag varieties in characteristic 0.

By combining Theorems 1.6.1 and 1.6.3, it easily follows that *the characteristic classes of any homogeneous principal G-bundle are algebraically trivial*, where G is any connected affine algebraic group. In fact, when G is reductive, the homogeneous G-bundles over A are exactly the *semi-stable G-bundles* with algebraically trivial characteristic classes, as shown by Mukai [Muk78] and Balaji-Biswas [BB02].

Finally, we introduce a notion of simplicity for a G-bundle $\pi : X \to Y$ over an arbitrary base, where G is a connected reductive group and $\mathrm{char}(k) = 0$. By analogy with Schur's lemma (a G-module is simple if and only if its equivariant automorphisms are just the scalars), we say that π is *simple* if any bundle automorphism is given by the action of a central element of G. When $G = \mathrm{GL}_n$, this is equivalent to the condition that the associated vector bundle $p : E := X \times^G k^n \to Y$ is simple, *i.e.*, its vector bundle endomorphisms are just scalars.

For an arbitrary G, a homogeneous G-bundle $\pi : X \to A$ is *simple if and only if X is a semi-abelian variety* (Proposition 7.3.3). In particular, there are very few simple homogeneous bundles. Thus, we introduce a weaker notion: we say that a G-bundle $\pi : X \to Y$ is *irreducible*, if the center of G has finite index in the group of bundle automorphisms. The irreducible homogeneous G-bundles are characterized by the following:

Proposition 1.6.4 *With the notation of Theorem 1.6.1, the following conditions are equivalent for a homogeneous bundle $\pi : X \to A$ under a connected reductive group G:*

(i) *π is irreducible.*

(ii) *H is diagonalizable and not contained in any proper Levi subgroup of G.*

(iii) *H is not contained in any proper parabolic subgroup of G.*

Remark 1.6.5 (i) Still assuming G to be connected and reductive, the structure of an arbitrary homogeneous G-bundle $\pi : X \to A$ may be somehow reduced to that of an irreducible homogeneous bundle under a Levi subgroup of G, as follows. Let $X = G \times^H \mathcal{G}$ as in Theorem 1.6.1 and identify H to its image in G (uniquely defined up to conjugation). Since H is commutative, it has a unique decomposition as $H = D \times U$, where D is a diagonalizable group and U a commutative unipotent group. By [BoTi71], there exists a parabolic subgroup $P \subset G$ and a Levi decomposition $P = R_u(P)L$ such that $U \subset R_u(P)$ and $D \subset L$; in particular, $H \subset P$. Moreover, we may assume that P is minimal for these properties (note however that they do not determine P uniquely). Then X has a reduction of structure group to the homogeneous P-bundle

$$\phi : Y := P \times^H \mathcal{G} \to A.$$

Moreover, ϕ factors through a principal bundle

$$\psi : Z := Y/R_u(P) \to A$$

under the connected reductive group $P/R_u(P) = L$, and this bundle is easily seen to be homogeneous by Theorem 1.6.3. The subgroup of L associated to ψ (via Theorem 1.6.1 again) is the image of H in $P/R_u(P)$, and hence is isomorphic to D; moreover, D is not contained in any proper parabolic subgroup of L, by our assumptions on P. Hence ψ is an irreducible L-bundle.

Thus, we may view homogeneous irreducible bundles as the 'building blocks' of homogeneous bundles.

(ii) An arbitrary algebraic subgroup H of a connected reductive algebraic group G is said to be *G-irreducible*, if H is not contained in any proper parabolic subgroup of G. Equivalently, H is reductive and not contained in any proper Levi subgroup of G.

For example, a subgroup H of $G = \mathrm{GL}_n$ is G-irreducible if and only if so is the representation of H in k^n. Returning to an arbitrary G, a subgroup H is G-irreducible if and only if its image in the quotient of G by its center Z is G/Z-irreducible; thus, we may assume that G *is semi-simple*. Then one easily shows that a *commutative* subgroup $H \subset G$ is G-irreducible if and only if $C_G(H)$ is finite; in particular, H itself is finite.

When $G = \mathrm{SL}_n$, any commutative G-irreducible subgroup is trivial, and this also holds when $G = \mathrm{Sp}_{2n}$. By the arguments of (i) above, it follows that any homogeneous bundle under SL_n or Sp_{2n} has a reduction of structure group to a homogeneous bundle under a Borel subgroup.

In contrast, when G is the projective linear group PGL_n, there are many commutative G-irreducible subgroups; in fact, they are exactly the images of the Heisenberg subgroups of GL_n acting via their standard irreducible representation (see *e.g.* [Mum66]).

For an arbitrary connected reductive group G, the classification of all commutative G-irreducible subgroups seems to be an open question. It turns out to be closely related to the torsion primes of G, as defined in [Ste75]. For example, if the order of a finite commutative subgroup H of G is not divisible by any torsion prime, then H is contained in a subtorus of G (see Corollary 2.25 in [loc. cit.]) and hence is not G-irreducible. The maximal elementary abelian p-subgroups of G are described in [Gri91] for all primes p; by the result quoted above, it suffices to consider the torsion primes.

Chapter 2

Proof of Chevalley's Theorem

In this chapter we shall present a proof of Chevalley's structure theorem (see Theorem 1.1.1) along the lines of Rosenlicht's proof in [Ros56].

We begin by proving several criteria for an algebraic group to be affine, that will be used in the course of the proof and in the next chapters. In particular, they imply that every connected algebraic group G has a largest connected normal affine subgroup G_{aff} (Lemma 2.1.2), and that G/Z_{red}^o is affine, where Z denotes the center of G, and Z_{red}^o its reduced neutral component (Corollary 2.1.7).

To prove Chevalley's theorem, we may thus assume that G_{aff} is trivial, and need to show that G is an abelian variety. But Z_{red}^o is nontrivial and contains no affine subgroup of positive dimension. If Z_{red}^o is an abelian variety, then the theorem follows almost immediately from the fact that every abelian subvariety of G is an almost direct factor (Corollary 2.2.4). We deduce this fact from a structure result for the action of an abelian variety on a smooth variety, which has independent interest (Theorem 2.2.2).

To complete the proof, it suffices to show that every non-complete algebraic group G contains an affine subgroup H of positive dimension. For this, the idea is to consider a completion X of G such that the G-action on itself by multiplication extends to an action on X, and to take for H the isotropy group of a point of the boundary. It is yet not known how to obtain such a completion at this stage (this will be done in Chapter 3, by using Chevalley's theorem; see Proposition 3.1.1). So we construct an equivariant completion in a weaker sense, namely, a complete variety birationally isomorphic to G, such that the induced rational action stabilizes a divisor (see Proposition 2.3.4). For this, we present some basic results on rational actions of algebraic groups.

2.1 Criteria for affineness

First, recall that an algebraic group is affine if and only if it is linear (see [Spr09, Theorem 2.3.7]). We shall also need the following:

Lemma 2.1.1 *Let G be an algebraic group and H a normal subgroup. Then G is affine if and only if H and G/H are affine.*

PROOF: If G is affine, then obviously H is affine; moreover, G/H is affine by [Spr09, Proposition 5.5.10]. To prove the converse, we embed H as a subgroup of some GL_n, and hence of SL_{n+1}. This realizes H as a closed subvariety of the affine space of $m \times m$ matrices, $\mathbb{M}_m$, where $m := n+1$. The group H acts naturally from the left on $\mathbb{M}_m$ (given the embedding) and from the right on G (via right multiplication). Then the associated bundle $E := G \times^H \mathbb{M}_m$ is a vector bundle over G/H (see Section 6.1 for details on associated bundles). We have the following commuting diagram:

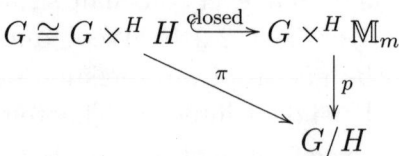

where π denotes the quotient morphism. As G/H is affine and the morphism $p : E \to G/H$ is affine (see for example [Har77, Exercise II.5.17]), E is affine. Hence G, being a closed subvariety of E, is affine.

Alternatively, it suffices to show that the morphism $\pi : G \to G/H$ is affine whenever H is affine. By arguing as in the proof of

[Har77, Theorem III.3.7], we reduce to showing that the functor π_* from the category of quasi-coherent sheaves on G to that of quasi-coherent sheaves on G/H, is exact. For this, we use the cartesian square

$$
\begin{array}{ccc}
G \times H & \xrightarrow{\ p_1\ } & G \\
\Big\downarrow{\scriptstyle m} & & \Big\downarrow{\scriptstyle \pi} \\
G & \xrightarrow{\ \pi\ } & G/H
\end{array}
$$

where p_1 denotes the projection, and m the multiplication. Then p_1 is clearly affine, and hence the functor $(p_1)_*$ is exact. Moreover, π is faithfully flat; hence we have a natural isomorphism $\pi^*(\pi_*\mathcal{F}) \cong (p_1)_*(m^*\mathcal{F})$ for any quasi-coherent sheaf $\mathcal{F}$ on G, by [loc. cit., Proposition III.9.3]. As m is faithfully flat, m^* is exact; it follows that $(p_1)_*m^*$ is exact. Using the exactness of π^*, this yields that π_* is exact, as required. $\qquad\square$

The above alternative argument is an example of *faithfully flat descent*; see, for example, [BLR90, Chapter 6] for further developments. We may now make a first step in the proof of Chevalley's theorem:

Lemma 2.1.2 *Any algebraic group G admits a largest connected affine normal subgroup G_{aff}. Moreover, $(G/G_{\mathrm{aff}})_{\mathrm{aff}}$ is trivial.*

PROOF: Let G_1 and G_2 be connected affine normal subgroups of G. Then $G_1 G_2$ is closed in G (indeed, consider the action of $G_1 \times G_2$ on G, where G_1 acts on the left and G_2 on the right. Then some orbit $G_1 g G_2$ is closed. But $G_1 g G_2 = G_1 G_2 g$ and hence $G_1 G_2$ is closed). Clearly $G_1 G_2$ is connected. Also, it is affine by Lemma 2.1.1, since $G_1 G_2/G_2 \cong G_1/G_1 \cap G_2$ is affine and G_2 is affine. Thus, if H is a connected affine normal subgroup of G of maximal possible dimension and K is any connected affine normal subgroup of G then by maximality of H we get $H = HK$, implying that $K \subset H$. This shows the existence of G_{aff}. The triviality of $(G/G_{\mathrm{aff}})_{\mathrm{aff}}$ follows again from Lemma 2.1.1. $\qquad\square$

In the next chapter, we shall use the generalization of Lemma 2.1.1 to the setting of an algebraic group G and a normal subgroup scheme H. This generalization follows readily from that lemma in view of the next result.

Lemma 2.1.3   *(i) For any group scheme H, we have*

$$H \text{ is affine} \Leftrightarrow H_{\mathrm{red}} \text{ is affine} \Leftrightarrow H^o_{\mathrm{red}} \text{ is affine.}$$

(ii) If H is a subgroup scheme of an algebraic group G, then

$$G/H \text{ is affine} \Leftrightarrow G/H_{\mathrm{red}} \text{ is affine} \Leftrightarrow G/H^o_{\mathrm{red}} \text{ is affine.}$$

PROOF: (i) The first equivalence holds for any scheme (see [Har77, Exercise III.3.1]). The second equivalence follows from the fact that H_{red} is the union of finitely many disjoint copies of H^o_{red}.

(ii) The quotient map $\pi : G \to G/H$ factors through a morphism $\varphi : G/H_{\mathrm{red}} \to G/H$ which is bijective, since its scheme-theoretic fibres are isomorphic to H/H_{red} and the latter is a finite scheme having a unique closed point. In fact, φ is also finite: indeed, we have a cartesian square as in the proof of Lemma 2.1.1:

$$
\begin{array}{ccc}
G \times H/H_{\mathrm{red}} & \xrightarrow{\;p_1\;} & G \\
{\scriptstyle m_1}\big\downarrow & & \big\downarrow{\scriptstyle \pi} \\
G/H_{\mathrm{red}} & \xrightarrow{\;\varphi\;} & G/H
\end{array}
$$

where p_1 denotes the projection, and m_1 comes from the multiplication $G \times H \to G$; since π is faithfully flat and p_1 is finite, φ is finite. In particular, φ is an affine morphism. Thus, if G/H is affine, then so is G/H_{red}. The converse follows from a theorem of Chevalley: if $f : X \to Y$ is a finite surjective morphism and if X is affine, then Y is affine (see [Har77, Exercise III.4.2]). This proves the first equivalence.

The second equivalence is checked similarly; here the morphism $G/H^o_{\mathrm{red}} \to G/H_{\mathrm{red}}$ is the quotient by the finite group $H_{\mathrm{red}}/H^o_{\mathrm{red}}$ acting via right multiplication, and hence this morphism is finite and surjective. $\square$

Definition 2.1.4 *An action of an algebraic group G on a variety X is* set-theoretic faithful *if every non-trivial $g \in G$ acts non-trivially. Equivalently, the set-theoretic kernel of the action is trivial.*

The action is scheme-theoretic faithful *if every non-trivial subgroup scheme of G acts non-trivially. Equivalently, the scheme-theoretic kernel of the action is trivial.*

A set-theoretic faithful action of G on X is scheme-theoretic faithful if and only if the induced action of the Lie algebra of G (by derivations of $\mathcal{O}_X$) is faithful. This condition always holds in characteristic 0, but fails in characteristic $p \geq 1$: the standard example is the action of $\mathbb{G}_m$ on $\mathbb{A}^1$ by $t \cdot z = t^p z$.

We shall only consider set-theoretic faithful actions in this chapter and the next one, and just call them faithful for simplicity. Scheme-theoretic faithfulness occurs in Chapter 4 and the subsequent chapters.

Lemma 2.1.5 *Let G be an algebraic group, and V a rational G-module (not necessarily finite-dimensional), i.e., every $v \in V$ sits in some finite-dimensional G-stable subspace on which G acts linearly and algebraically. If the G-action on V is faithful, then G is affine.*

PROOF: By assumption, $V = \bigcup_{i \in I} V_i$, where the V_i are finite-dimensional rational G-modules. Let K_i denote the set-theoretic kernel of the G-action on V_i. Then K_i is a normal subgroup of G, and the set-theoretic intersection $\bigcap_{i \in I} K_i$ is trivial. Since the underlying topological space of G is noetherian, there exists a finite subset $J \subset I$ such that $\bigcap_{j \in J} K_j$ is trivial. Thus, G acts faithfully on $W := \sum_{j \in J} V_j$, a finite-dimensional submodule of V. Let H denote the scheme-theoretic kernel of the representation $G \to \mathrm{GL}(W)$. Then H_{red} is trivial, and hence the group scheme H is finite; moreover, G/H is affine as a subgroup of $\mathrm{GL}(W)$. Hence G is affine by Lemmas 2.1.1 and 2.1.3. $\square$

Proposition 2.1.6 *Let G be an algebraic group acting faithfully on a variety X with a fixed point x. Then G is affine.*

PROOF: By Lemma 2.1.5, it suffices to construct a faithful rational representation of G. Since G fixes x, it acts linearly (but not rationally) on the local ring $\mathcal{O}_{X,x}$, stabilizing the ideals $\mathfrak{m}_x^n$ for all $n > 0$, where $\mathfrak{m}_x$ denotes the maximal ideal in $\mathcal{O}_{X,x}$. Thus, G acts linearly on the finite dimensional vector space $\mathcal{O}_{X,x}/\mathfrak{m}_x^n$ for any $n \geq 1$.

We now show that this representation is rational. Indeed, the action morphism

$$\alpha : G \times X \to X, \quad (g,x) \mapsto g \cdot x$$

induces a homomorphism of local rings

$$\alpha^\# : \mathcal{O}_{X,x} \to \mathcal{O}_{G \times X,(g,x)}.$$

But $\mathcal{O}_{G \times X,(g,x)}$ is the localization of $\mathcal{O}_{G,g} \otimes \mathcal{O}_{X,x}$ at the maximal ideal of (g,x), *i.e.*, at $\mathfrak{m}_g \otimes \mathcal{O}_{X,x} + \mathcal{O}_{G,g} \otimes \mathfrak{m}_x$. Moreover, we have

$$\mathcal{O}_{G \times X,(g,x)}/\mathfrak{m}_x^n \mathcal{O}_{G \times X,(g,x)} \cong \mathcal{O}_{G,g} \otimes \mathcal{O}_{X,x}/\mathfrak{m}_x^n$$

since the right-hand side is already a local ring. Thus, for each $n \geq 1$ we obtain a homomorphism

$$\alpha_n^\# : \mathcal{O}_{X,x}/\mathfrak{m}_x^n \to \mathcal{O}_{G,g} \otimes \mathcal{O}_{X,x}/\mathfrak{m}_x^n.$$

This means that the matrix coefficients of the G-action on $\mathcal{O}_{X,x}/\mathfrak{m}_x^n$ lie in $\mathcal{O}_{G,g}$ for any $g \in G$, and hence in $\mathcal{O}(G)$ as required.

Next, let K_n denote the set-theoretic kernel of the representation

$$G \to \mathrm{GL}(\mathcal{O}_{X,x}/\mathfrak{m}_x^n).$$

Clearly, K_n is contained in K_{n-1}. So we have the following decreasing sequence of subgroups of G:

$$K_1 \supseteq \cdots \supseteq K_n \supseteq K_{n+1} \supseteq \cdots$$

which must stabilize at some stage, say n_0. Thus, K_{n_0} acts trivially on $\mathcal{O}_{X,x}/\mathfrak{m}_x^n$ for all $n \geq n_0$. We now show that K_{n_0} is trivial, giving us the representation of G that we seek. By Krull's intersection theorem we know that

$$\bigcap_{n \geq 1} \mathfrak{m}_x^n = 0.$$

Hence K_{n_0} acts trivially also on $\mathcal{O}_{X,x}$, and on its field of fractions – the function field $k(X)$. As the local ring $\mathcal{O}_{X,y}$ is contained in $k(X)$ for every $y \in X$, we get that K_{n_0} acts trivially on $\mathcal{O}_{X,y}$ which implies that K_{n_0} fixes all points in X. By the faithfulness assumption, we see that K_{n_0} is trivial. $\qquad\square$

As a direct consequence, we obtain:

Corollary 2.1.7 *Let G be a connected algebraic group, and $Z = Z(G)$ its scheme-theoretic center. Then G/Z, G/Z_{red} and G/Z_{red}^o are affine.*

PROOF: G acts by conjugation on itself, and the set-theoretic kernel of this action is Z_{red}. Hence G/Z_{red} acts faithfully on the variety G. The neutral element e_G is a fixed point for this action. Hence, applying Proposition 2.1.6 yields that G/Z_{red} is affine. The affineness of G/Z and of G/Z_{red}^o follows in view of Lemma 2.1.3. $\qquad\square$

Remark 2.1.8 The above statement does not extend to arbitrary algebraic groups. For example, let E be an elliptic curve, and G the semi-direct product of E with the group of order 2 generated by the multiplication $(-1)_E$. Then one checks that $Z = E_2$ (the kernel of 2_E) and hence G/Z is not affine.

Another direct consequence of Lemma 2.1.3 and of Proposition 2.1.6 is the following:

Corollary 2.1.9 *Let X be a variety equipped with a faithful action of an algebraic group G. Then the isotropy subgroup scheme of any $x \in X$ is affine.* $\qquad\square$

2.2 Actions of abelian varieties

We first record the following observation:

Proposition 2.2.1 *Let A be an abelian variety acting faithfully on a variety X. Then the isotropy subgroup scheme A_x is finite for every $x \in X$.*

(Indeed, A_x is proper, but also affine by Corollary 2.1.9).   □

The next theorem is a variant of a result of Rosenlicht (see [Ros56, Theorem 14]):

Theorem 2.2.2 *Let A be an abelian variety acting faithfully on a smooth variety X. Then there exists a morphism $\phi : X \to A$ which is equivariant in the following sense:*

$$\phi(ax) = \phi(x) + na \quad \text{for all } x \in X, \quad a \in A,$$

where n is a positive integer.

Note that the action of A on itself via $a \cdot b := b + na$ is just the action on the quotient A/A_n by translation, where A_n denotes the n-torsion subgroup scheme of A. So the above theorem asserts the existence of an equivariant morphism $\phi : X \to A/A_n$ for some $n > 0$; in other words, there is an equivariant isomorphism

$$X \cong A \times^{A_n} Y$$

where Y is a closed subscheme of X, stable under A_n (the fibre of ϕ at the origin of A).

A result of Nishi and Matsumura states more generally that, if X is a smooth variety equipped with a faithful action of a connected algebraic group G then there exists an equivariant morphism $\phi : X \to G/H$, where $H \subset G$ is the pre-image of some $A_n \subset A = G/G_{\text{aff}}$. Equivalently, $X = G \times^H Y$ for some closed H-stable subscheme $Y \subset X$; see [Mat63], and also [Bri10b] for another proof and further developments.

Before proving Theorem 2.2.2, we show that the assumption of smoothness cannot be suppressed by presenting an example due to Raynaud ([Ray70, XIII 3.1]).

Example 2.2.3 Let E be an elliptic curve, and $p_0 \in E$ a point of infinite order. Define an equivalence relation $\sim$ on $E \times \mathbb{P}^1$ by setting

$$(p, 0) \sim (p + p_0, \infty).$$

Let $X := E \times \mathbb{P}^1 / \sim$. Equivalently, X is obtained by gluing together $E \times \{0\}$ and $E \times \{\infty\}$ in $E \times \mathbb{P}^1$ via the translation τ_{p_0}. Then X is a variety (as follows e.g. from [Fer03, Theorem 5.4]); in fact, X is a complete surface. Moreover, the natural map $\eta : E \times \mathbb{P}^1 \to X$ is the normalization. The elliptic curve E acts faithfully on X via its action on $E \times \mathbb{P}^1$ by translation on E. We claim that there is no such morphism $\phi : X \to E$ as in the theorem. For, if such an E-equivariant morphism does exist, then it yields a morphism $\psi : E \times \mathbb{P}^1 \to E$ such that the following diagram commutes

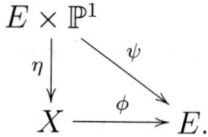

Then the morphism $\mathbb{P}^1 \to E$, $z \mapsto \psi(p, z)$ must be constant for any $p \in E$. By the rigidity lemma for complete varieties applied to ψ (see e.g. [Mil86, Theorem 2.1]) we get $\psi(p, z) = \gamma(p)$ for some morphism $\gamma : E \to E$. By the assumption on ϕ we obtain

$$\gamma(p + p_0) = \gamma(p) + np_0 \text{ for all } p \in E, z \in \mathbb{P}^1,$$

for some $n > 0$. Also, $\psi(p, 0) = \psi(p + p_0, \infty)$ so that $\gamma(p) = \gamma(p + p_0)$. Thus, $np_0 = 0$ thereby contradicting that p_0 is of infinite order. $\qquad\square$

PROOF OF THEOREM 2.2.2: Let $x_0 \in X$ and let $Y := A \cdot x_0$ be the orbit through x_0. By Proposition 2.2.1, we have $Y \cong A/F$ where F is a finite subgroup scheme of A. We now claim that there exists a line bundle L on X such that $L \mid_Y$ is ample. Indeed, let $E \subset Y$ be an ample irreducible divisor (we can choose one such because Y is an abelian variety). Then there exists an effective divisor, $D \subset X$, such that $D \cap Y \supseteq E$, and D does not contain Y. (Indeed we can pick a local equation g of E at some point x, and lift it to a rational function f on X which is defined at x. The closure in X of the zero scheme of f is an effective Weil divisor D, and hence a Cartier divisor since X is smooth. As E is irreducible, it is contained in an irreducible component of D. Then D is the required effective divisor.) Let $L := \mathcal{O}_X(D)$. Then $L|_Y = \mathcal{O}_Y(E + F)$ where F is an effective divisor. Since A is an abelian variety, it follows that F is numerically effective. (Indeed, given an irreducible curve C in A, we may find $x \in A$ such that $\tau_x(C)$ is not contained in the support of D. Then the intersection number $D \cdot \tau_x(C)$ is non-negative. But $D \cdot \tau_x(C) = D \cdot C$ as $\tau_x(C)$ is algebraically equivalent to C. Hence $D \cdot C \geq 0$ as required). Thus, $L|_Y$ is ample.

Let $\mathcal{L} := \alpha^*(L) \otimes p_2^*(L^{-1})$ where $\alpha : A \times X \to X$ denotes the action and $p_2 : A \times X \to X$ the projection. Then $\mathcal{L}_{0 \times X}$ is the trivial bundle. Hence $\mathcal{L}$ defines a morphism

$$\varphi : X \to \operatorname{Pic}(A), \quad x \mapsto \mathcal{L}|_{A \times x}$$

(see [BLR90, §8.2], [Mum08, §III.13]).

For any $x \in X$, consider the orbit map $\alpha_x : A \to X$, $a \mapsto a \cdot x$. Then α_x is finite by Proposition 2.2.1, and $\varphi(x)$ is the class of $\alpha_x^*(L)$ in $\operatorname{Pic}(A)$, since $p_2^*(L)|_{A \times x}$ is trivial. Thus, $\varphi(x_0)$ is ample. Also, $\varphi(X)$ is contained in a connected component of $\operatorname{Pic}(A)$, i.e., in a coset of $\operatorname{Pic}^\circ(A) = \hat{A}$. It follows that $\varphi(x)$ is ample for any $x \in X$. Since α_x is A-equivariant, we see that φ is also equivariant with respect to the given action on X and the action on $\operatorname{Pic}(A)$ by translation on A. The latter action is transitive on any ample coset, by [Mil86, Proposition 10.1]. Thus we get a morphism

$$\varphi : X \to A \cdot \varphi(x_0) \cong A/K(\alpha_{x_0}^*(L)),$$

where for any line bundle M on A, we denote by $K(M)$ the scheme-theoretic kernel of the polarization homomorphism

$$\varphi_M : A \to \hat{A}, \quad a \mapsto \tau_a^*(M) \otimes M^{-1}.$$

Since $\alpha_{x_0}^*(L)$ is ample, the group scheme $K(\alpha_{x_0}^*(L))$ is finite (see [Mil86, Proposition 9.1]), and hence is contained in A_n for some $n > 0$. Thus, φ yields an A-equivariant morphism $X \to A/A_n$, and hence a morphism $\phi : X \to A$ as in the statement. $\square$

From the above theorem we arrive at a generalization of the Poincaré complete reducibility theorem (see for example [Mil86, Proposition 12.1]) to the setting of connected algebraic groups.

Corollary 2.2.4 *Let A be an abelian subvariety of a connected algebraic group G. Then A is contained in the center of G, and there exists a connected normal subgroup H of G such that $G = AH$ and $A \cap H$ is finite.*

PROOF: The first assertion follows, for example, from Corollary 2.1.7. Applying Theorem 2.2.2 to the action of A on G by multiplication, we get a morphism $\phi : G \to A$ such that $\phi(ag) = \phi(g) + na$ for some positive integer n. We may assume without loss of generality that $\phi(e_G) = 0_A$; then ϕ is a group homomorphism (see [Mil86, Corollary 3.6]). As $\phi|_A$ is just the multiplication n_A, it is an isogeny. Thus, we get that $G = A \ker(\phi)$ and $A \cap \ker(\phi) = \ker(n_A) = A_n$ is finite. Now, let $H := (\ker \phi)^o_{\mathrm{red}}$; then we still have $G = AH$ and $H \cap A$ is finite. □

The above corollary is due to Rosenlicht (see [Ros56, Corollary, p. 434]); a subgroup $H \subset G$ as in the statement shall be called a *quasi-complement* to A in G.

2.3 Rational actions of algebraic groups

Definition 2.3.1 *We denote by $\mathrm{Bir}(X)$ the group of birational automorphisms of a variety X; this is also the group of k-automorphisms of the function field $k(X)$.*

A rational action of an algebraic group G on X is a group homomorphism $\rho : G \to \mathrm{Bir}(X)$ such that the map $\alpha : G \times X \dashrightarrow X$, $(g, x) \mapsto \rho(g)x =: g \cdot x$ is defined and regular on a dense open subset of X. We say that $g \cdot x$ is defined if the rational map α is defined at $(g, x) \in G \times X$.

A rational action ρ is faithful if its kernel is trivial.

A point $x \in X$ is fixed by this action, if there exists a dense open subset $V \subset G$ such that $g \cdot x$ is defined and equals x for any $g \in V$.

We now obtain a version of Proposition 2.1.6 for rational actions:

Proposition 2.3.2 *Let G be an algebraic group acting rationally and faithfully on a variety X with a fixed point x. Then G is affine.*

PROOF: We adapt the argument of Proposition 2.1.6. We may assume that G is connected. The dense open subset V of G (as in Definition 2.3.1) stabilizes x, hence the local ring $\mathcal{O}_{X,x} \subset k(X)$ is also stabilized by V. Since V generates the group G, we see that G itself stabilizes $\mathcal{O}_{X,x}$. Thus G stabilizes the unique maximal ideal $\mathfrak{m}_x$ of $\mathcal{O}_{X,x}$, and its powers $\mathfrak{m}_x^n$; hence G acts linearly on the quotients $\mathcal{O}_{X,x}/\mathfrak{m}_x^n$.

We show that the matrix coefficients of the action of G on $\mathcal{O}_{X,x}/\mathfrak{m}_x^n$ are in $k(G)$. Indeed, the co-action map

$$\alpha^\# : k(X) \to k(G \times X)$$

restricts to a homomorphism of local rings

$$\alpha^\# : \mathcal{O}_{X,x} \to \mathcal{O}_{U,V \times x} = \mathcal{O}_{G \times X, G \times x}.$$

The right-hand side is the localization of $k(G) \otimes \mathcal{O}_{X,x}$ at the maximal ideal $k(G) \otimes \mathfrak{m}_x$. Moreover, for the maximal ideal $\mathfrak{m}$ of $\mathcal{O}_{G \times X, G \times x}$ we have

$$\mathcal{O}_{G \times X, G \times x}/\mathfrak{m}^n \cong k(G) \otimes \mathcal{O}_{X,x}/\mathfrak{m}_x^n$$

for all $n \geq 1$. Thus, we obtain homomorphisms

$$\alpha_n^\# : \mathcal{O}_{X,x}/\mathfrak{m}_x^n \to k(G) \otimes \mathcal{O}_{X,x}/\mathfrak{m}_x^n.$$

This is equivalent to giving a rational homomorphism

$$\phi : G \dashrightarrow \mathrm{GL}(\mathcal{O}_{X,x}/\mathfrak{m}_x^n) =: \mathrm{GL}_N.$$

Therefore, there exists a dense open subset W of G on which ϕ is regular. Let Γ_W be the graph of this morphism and let Γ denote the closure of Γ_W in $G \times \mathrm{GL}_N$. Then Γ is a subgroup of $G \times \mathrm{GL}_N$, and we have a commuting diagram of group homomorphisms

The map $p_1 : \Gamma \to G$ is birational since ϕ is a rational map, hence an isomorphism since it is a group homomorphism. Thus, from the commutativity of the above diagram we conclude that the map ϕ is regular. $\qquad\square$

We also record a result of Rosenlicht (see [Ros56, Lemma, p. 403]):

Lemma 2.3.3 *Let X be a variety equipped with a birational action of a connected algebraic group G, and let $x \in X$, $g, h \in G$. If $h \cdot x$ and $g \cdot (h \cdot x)$ are defined, then so is $gh \cdot x$, and we have $gh \cdot x = g \cdot (h \cdot x)$.*

PROOF: We shall use the following observation : let $f : Y \dashrightarrow Z$ be a rational map of varieties, V the largest open subset of Y on which f is defined, and $\Gamma \subset Y \times Z$ the graph of f (*i.e.*, the closure of the graph of the morphism $f \mid_V$). Then f is defined at a point $y \in Y$ if and only if the projection $p_1 : \Gamma \to Y$ is an isomorphism above some neighborhood of y.

Now consider the graph of the rational map

$$G \times G \times X \dashrightarrow X \times X \times X, \quad (u, v, z) \mapsto (v \cdot z, u \cdot (v \cdot z), uv \cdot z).$$

This graph Γ sits in $G \times G \times X \times X \times X \times X$; actually, in $G \times G \times X \times X \times \Delta(X)$ where $\Delta(X)$ denotes the diagonal (as $u \cdot (v \cdot z) = uv \cdot z$ on a dense open subset of $G \times G \times X$). Since the rational map $(u, v, z) \mapsto (v \cdot z, (u \cdot (v \cdot z)))$ is defined at (g, h, x), the projection $\Gamma \to G \times G \times X$ to the first three factors is an isomorphism above a neighborhood of (g, h, x). It follows that the rational map $(u, v, z) \mapsto uv \cdot z$ is defined at (g, h, x), with value $g \cdot (h \cdot x)$ at that point. $\qquad\square$

Next, we present another result of Rosenlicht, on "equivariant completion" of rational actions (see [Ros56, Theorem 15]):

Proposition 2.3.4 *Let X_0 be a non-complete variety equipped with a regular action of a connected algebraic group G. Then X_0 is equivariantly birationally isomorphic to a complete normal variety*

X equipped with a rational action α of G, and containing an irreducible divisor D such that α restricts to a rational action of G on D.

PROOF: By a theorem of Nagata, we may embed X_0 as an open subset of a complete variety X. We may assume that $X \setminus X_0$ has pure codimension 1 by blowing-up X along $X \setminus X_0$; then replacing X with its normalization, we may also assume that X is normal. Let $\alpha : G \times X \dashrightarrow X$ be the rational action arising from the given action on X_0, and choose an irreducible component E of $X \setminus X_0$. Then α is defined on $G \times E$ (since the latter is a divisor of the normal variety $G \times X$, and the target X is complete; apply the valuative criterion of properness [Har77, Theorem II.4.7]). Moreover, the restriction $G \times E \dashrightarrow X$ is not dominant. For otherwise, $g \cdot x$ is defined and lies in X_0 for any general point $(g, x) \in G \times E$; thus, $g^{-1} \cdot (g \cdot x)$ is defined, and equals x by Lemma 2.3.3. Hence $x \in X_0$, a contradiction.

By Lemma 2.3.5 below, there exists a complete variety X' and a birational morphism $\varphi : X' \to X$ such that the induced rational map $G \times X \dashrightarrow X'$ sends $G \times E$ to a divisor; we may further assume that X' is normal. Then X' is equipped with a rational action $\alpha' : G \times X' \dashrightarrow X'$ and with an irreducible divisor E' (birationally isomorphic to E) such that α' sends $G \times E'$ to an irreducible divisor D'.

We claim that X', α' and D' satisfy the assertions of the proposition. Indeed, $g \cdot x$ is defined for a general point $(g, x) \in G \times D'$. We may further write $x = h \cdot y$ for general $(h, y) \in G \times E'$. Then $g \cdot x = g \cdot (h \cdot y)$ equals $gh \cdot y$ by Lemma 2.3.3 again; Thus, $g \cdot x$ lies in D', that is, α' sends $G \times D'$ to D'. Clearly, the resulting map $G \times D' \dashrightarrow D'$ is rational and yields a homomorphism $G \to \mathrm{Bir}(D')$. $\square$

Lemma 2.3.5 Let X be a normal variety, $D \subset X$ an irreducible divisor, Y a complete variety, and $f : X \dashrightarrow Y$ a dominant rational map. Then there exist a complete variety Y' and a birational morphism $\varphi : Y' \to Y$ such that the induced rational map $f' : X \dashrightarrow Y'$

restricts to a rational map $f'|_D \colon D \dashrightarrow Y'$ *with image* Y' *or a divisor.*

PROOF: We may assume that $f|_D$ is not dominant. We shall view the function field $k(Y)$ as a subfield of $k(X)$ via the comorphism $f^\#$. Let v denote the valuation of $k(X)$ associated with the irreducible divisor D, so that $v(f)$ is the order of the zero or pole of $f \in k(X)^*$ along D. Then v is discrete and its residue field k_v is the function field of D. Next, let w denote the restriction of v to $k(Y)$. Then w is also a discrete valuation, and is nontrivial since $f|_D$ is not dominant. Its residue field k_w is contained in k_v, and we have the inequality of transcendence degrees

$$\text{tr.deg.}k_v/k_w \leq \text{tr.deg.}k(X)/k(Y).$$

(Indeed, let $g_1, \ldots, g_t \in k_v$ be algebraically independent over k_w, and choose representatives $h_1, \ldots, h_t$ of $g_1, \ldots, g_t$ in the valuation ring $\mathcal{O}_v$. Then one easily checks that $h_1, \ldots, h_t$ are algebraically independent over $\mathcal{O}_w$, and hence over $k(Y)$). But $\text{tr.deg.}k(X)/k(Y) = m - n$, where $m := \dim(X)$ and $n := \dim(Y)$. Also, $\text{tr.deg.}k_v/k = m - 1$. Thus,

$$\text{tr.deg.}k_w/k = \text{tr.deg.}k_v/k - \text{tr.deg.}k_v/k_w \geq m - 1 - (m-n) = n - 1.$$

So we may choose functions $f_1, \ldots, f_{n-1} \in \mathcal{O}_w$ (the local ring of w) with algebraically independent images in k_w. Let Y' be the graph of the rational map

$$Y \dashrightarrow \mathbb{P}^{n-1}, \quad y \mapsto [1 : f_1(y) : \cdots : f_{n-1}(y)].$$

Then Y' is a complete variety equipped with a birational morphism $p_1 \colon Y' \to Y$ and with a morphism $p_2 \colon Y' \to \mathbb{P}^{n-1}$. Thus, f lifts to a rational map $f' \colon X \dashrightarrow Y'$; moreover, the composition $p_2 \circ f' \colon X \dashrightarrow \mathbb{P}^{n-1}$ restricts to a dominant rational map $D \dashrightarrow \mathbb{P}^{n-1}$, by construction of $f_1, \ldots, f_{n-1}$. Hence f' maps D to a divisor as required. $\qquad \square$

The final ingredient of the proof of Chevalley's theorem is the following lemma, also due to Rosenlicht (see [Ros56, Lemma 1, p. 437]).

Lemma 2.3.6 *Any non-complete algebraic group has an affine subgroup of positive dimension.*

PROOF: We may assume that G is connected. Applying Proposition 2.3.4 to the variety G where G acts by multiplication, we obtain a complete normal variety X equipped with a rational action α and with an irreducible divisor D, stable by this action.

We claim that there exists an open subset V of G and a point $x_0 \in D$ such that $g \cdot x_0$ and $g^{-1} \cdot (g \cdot x_0)$ are defined for any $g \in V$. Indeed, let $U \subset G \times X$ be the largest open subset on which α is defined; then U contains a dense open subset of $G \times D$, as seen in the proof of Proposition 2.3.4. Since G acts rationally on D, the morphism $(G \times D) \cap U \to G \times D$, $(g, x) \mapsto (g^{-1}, g \cdot x)$ is dominant, and hence the subset $\{(g, x) \in G \times D \mid (g, x) \in U, (g^{-1}, g \cdot x) \in U\}$ is open and dense in $G \times D$. This implies our claim.

Choose $g_0 \in V$ and denote by F the (set-theoretic) fiber at g_0 of the "orbit map" $V \to D$, $g \mapsto g \cdot x_0$. Then each irreducible component of F has dimension ≥ 1, since $\dim(V) = \dim(G) = \dim(D) + 1$. Note that $g \cdot x_0$ is defined and equals $g_0 \cdot x_0$ for any $g \in F$. Thus, $g_o^{-1} \cdot (g \cdot x_0) = g_0^{-1} \cdot (g_0 \cdot x_0)$ is defined and equals x_0. By Lemma 2.3.3, it follows that $g_0^{-1} g \cdot x_0$ is defined and equals x_0. In other words, $g_0^{-1} F$ fixes x_0. Likewise, $g^{-1} \cdot (g_0 \cdot x_0) = g^{-1} \cdot (g \cdot x_0) = x_0$ for any $g \in F$, i.e., $F g_0^{-1}$ fixes x_0.

Next, choose an irreducible component C of F containing g_0. Then $g_0^{-1} C$ is a locally closed subvariety of G containing e_G. Thus, the subgroup H of G generated by $g_0^{-1} C$ is closed in G and of positive dimension. Moreover, H fixes x_0 in view of Lemma 2.3.3 again. So H is affine by Proposition 2.3.2.   □

Remark 2.3.7 We present an alternative proof of the above proposition, based on a theorem of Weil: any variety equipped with a birational action of a connected algebraic group G is equivariantly birationally isomorphic to a variety equipped with a regular action of G.

By that theorem, there exists a variety E equivariantly birationally isomorphic to D and on which G acts regularly. If G fixes E pointwise, then it has a fixed point in D, and hence is affine.

Thus, we may assume that G acts non-trivially on E. Since the isotropy group of each point of E has positive dimension, G strictly contains a non-trivial connected subgroup H.

If H is non-complete, then we conclude by induction on the dimension of G (the case where $\dim(G) = 1$ is easy, since G is then a non-complete irreducible curve and hence is affine). But if H is complete, then it admits a quasi-complement K by Corollary 2.2.4. Then K is non-complete (since so is G), and $K \neq G$. Thus, we may again conclude by induction on $\dim(G)$.

PROOF OF THEOREM 1.1.1: We provide details for the argument outlined at the beginning of this chapter. By Lemma 2.1.2, it suffices to show that G/G_{aff} is an abelian variety. Thus, we may assume that every connected normal affine subgroup of G is trivial. We now show that G is complete.

Consider the center Z of G. If Z is not complete, then it contains a connected affine subgroup of positive dimension by Lemma 2.3.6; such a subgroup being normal, this contradicts our assumption on G. Hence Z is complete, and its reduced neutral component Z^o_{red} is an abelian variety. Let $H \subset G$ be a quasi-complement to Z^o_{red} in G, as in Corollary 2.2.4. Then $G = Z^o_{\mathrm{red}}H$ and $Z^o_{\mathrm{red}} \cap H$ is finite. Also, $H/(Z^o_{\mathrm{red}} \cap H) \cong G/Z^o_{\mathrm{red}}$ is affine by Corollary 2.1.7. Thus, H is affine as well (Lemma 2.1.1). So H is trivial by our assumption on G, *i.e.*, $G = Z$ is complete.

Chapter 3

Applications and developments of Chevalley's theorem

We begin this chapter by deriving some applications of Chevalley's theorem (Theorem 1.1.1) and of the structure of actions of abelian varieties on smooth varieties (Theorem 2.2.2). Then we present a proof of the "dual" statement of Chevalley's theorem (Theorem 1.2.1). Finally, we combine both structure theorems to obtain the Rosenlicht decomposition (Theorem 1.2.4) and some further developments on the structure of algebraic groups.

3.1 Some applications

Proposition 3.1.1 *Let G be a connected algebraic group. Then:*

(i) G_{aff} is the largest connected affine subgroup of G.

(ii) $G = Z^o_{\mathrm{red}} G_{\mathrm{aff}} = Z G_{\mathrm{aff}}$ where $Z = Z(G)$ denotes the scheme-theoretic center of G. Moreover, $Z \cap G_{\mathrm{aff}} = Z(G_{\mathrm{aff}})$ (the scheme-theoretic center of G_{aff}). In particular, G is commutative if and only if G_{aff} is so.

(iii) The quotient homomorphism

$$\alpha : G \to G/G_{\mathrm{aff}} =: A$$

is a Zariski locally trivial principal G_{aff}-bundle. Further, α is the Albanese morphism of G.

(iv) G admits a $G \times G$-equivariant completion by a normal projective variety.

PROOF: (i) Let H be a connected affine subgroup of G. Then $\alpha(H)$ is a connected affine subgroup of G/G_{aff}. Hence by the completeness of G/G_{aff}, $\alpha(H)$ is trivial. Thus $H \subset G_{\mathrm{aff}}$ as required.

(ii) Since G/G_{aff} is complete, so also is its quotient $G/Z^o_{\mathrm{red}}G_{\mathrm{aff}}$. However, G/Z^o_{red} is affine by Corollary 2.1.7, so that the quotient $G/Z^o_{\mathrm{red}}G_{\mathrm{aff}}$ is also affine. Hence $G/Z^o_{\mathrm{red}}G_{\mathrm{aff}}$ is trivial, that is, $G = Z^o_{\mathrm{red}}G_{\mathrm{aff}}$. It follows that $G = ZG_{\mathrm{aff}}$, and hence $Z(G_{\mathrm{aff}}) \subset Z \cap G_{\mathrm{aff}}$. But the opposite inclusion is obvious; this completes the proof.

(iii) Let $f : G \to B$ be a morphism to an abelian variety. Then f is the composite of a group homomorphism $\varphi : G \to B$ and a translation of B (see [Mil86, Corollary 3.6]). The group $\varphi(G_{\mathrm{aff}})$ is trivial (being both affine and complete), and hence φ factors through a unique homomorphism $A \to B$. Thus, f also factors through a unique morphism $A \to B$. This shows that α is the Albanese morphism of G.

To show that the G_{aff}-bundle α is locally trivial, we reduce to the case where G_{aff} is solvable, as follows. Let B be a Borel subgroup of G_{aff}. Then ZB is a subgroup scheme of G. By (ii) above, we know that $\alpha \mid_Z$ is onto, hence so also is $\alpha \mid_{ZB}$. Clearly,

$$B \subset \ker(\alpha \mid_{ZB}) = ZB \cap G_{\mathrm{aff}} = (Z \cap G_{\mathrm{aff}})B = Z(G_{\mathrm{aff}})B.$$

Moreover, $Z(G_{\mathrm{aff}}) \subset B$ by Lemma 3.1.2 below. This yields the equality

$$B = \ker \alpha \mid_{ZB} = ZB \cap G_{\mathrm{aff}}. \tag{3.1}$$

Thus, $\alpha \mid_{ZB}$ is a principal bundle under the connected solvable affine algebraic group B. By [Ser01, Proposition 14], any such bundle has

(Zariski) local sections. On the other hand, it follows from (3.1) that

$$G_{\mathrm{aff}} \times^{B} ZB = G_{\mathrm{aff}} \times^{G_{\mathrm{aff}} \cap ZB} ZB \cong G_{\mathrm{aff}} ZB = G$$

and also that

$$ZB/B = ZB/ZB \cap G_{\mathrm{aff}} = ZBG_{\mathrm{aff}}/G_{\mathrm{aff}} = G/G_{\mathrm{aff}} = A.$$

Thus, the extension of structure group of the principal B-bundle $\alpha \mid_{ZB}: ZB \to ZB/B$ to G_{aff}, is just $\alpha: G \to A$. In particular, each local section of $\alpha \mid_{ZB}$ yields a local section of α. Hence the latter G_{aff}-bundle is locally trivial, as required.

(iv) We may view G_{aff} as a closed subgroup of some GL_n. Then

$$G_{\mathrm{aff}} \subset \mathbb{M}_n \subset \mathbb{P}(\mathbb{M}_n \oplus k).$$

Let $\overline{G_{\mathrm{aff}}}$ be the closure of G_{aff} in $\mathbb{P}(\mathbb{M}_n \oplus k)$. The action of $G_{\mathrm{aff}} \times G_{\mathrm{aff}}$ on G_{aff} by left and right multiplication extends to $\overline{G_{\mathrm{aff}}}$. Taking the normalization of $\overline{G_{\mathrm{aff}}}$, we get a normal projective $G_{\mathrm{aff}} \times G_{\mathrm{aff}}$-equivariant completion X of G_{aff}. We have the following diagram

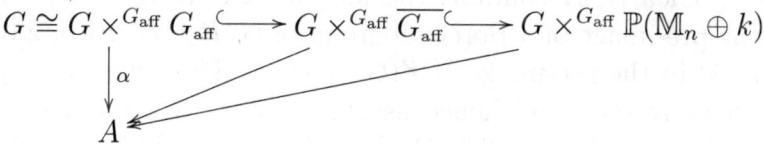

Further, $G \times^{G_{\mathrm{aff}}} \mathbb{P}(\mathbb{M}_n \oplus k)$ is a projective variety (the projective completion of the vector bundle $G \times^{G_{\mathrm{aff}}} \mathbb{M}_n$ over the projective variety A); thus, so are $G \times^{G_{\mathrm{aff}}} \overline{G_{\mathrm{aff}}}$ and its normalization $G \times^{G_{\mathrm{aff}}} X$.

We now show that the map $G \hookrightarrow G \times^{G_{\mathrm{aff}}} X$ is $G \times G$-equivariant. One easily checks that the map $G \to G \times G$, $g \mapsto (e_G, g)$ induces an isomorphism

$$G/G_{\mathrm{aff}} \to (G \times G)/(G_{\mathrm{aff}} \times G_{\mathrm{aff}})Z,$$

where Z is embedded in $G \times G$ via $z \mapsto (z, z)$. Moreover,

$$
\begin{aligned}
(G_{\mathrm{aff}} \times G_{\mathrm{aff}})Z &\cong (G_{\mathrm{aff}} \times G_{\mathrm{aff}} \times Z)/(G_{\mathrm{aff}} \times G_{\mathrm{aff}}) \cap Z \\
&\cong (G_{\mathrm{aff}} \times G_{\mathrm{aff}} \times Z)/Z(G_{\mathrm{aff}}),
\end{aligned}
$$

where $Z(G_{\mathrm{aff}})$ is embedded in $G_{\mathrm{aff}} \times G_{\mathrm{aff}} \times Z$ via $z \mapsto (z, z, z)$. Thus,

$$G \times^{G_{\mathrm{aff}}} X \cong G \times G \times^{(G_{\mathrm{aff}} \times G_{\mathrm{aff}})Z} X,$$

where $(G_{\mathrm{aff}} \times G_{\mathrm{aff}})Z$ acts on $G \times G$ by right multiplication, and on X via the given action of $G_{\mathrm{aff}} \times G_{\mathrm{aff}}$ and the trivial action of Z (this defines an action of the product $G_{\mathrm{aff}} \times G_{\mathrm{aff}} \times Z$ which restricts to the trivial action of $Z(G_{\mathrm{aff}})$; so this indeed defines an action of $(G_{\mathrm{aff}} \times G_{\mathrm{aff}})Z$). Now it is easy to see that the right-hand side has an action of $G \times G$, extending the action on the open subset

$$G \cong G \times^{G_{\mathrm{aff}}} G_{\mathrm{aff}} \cong (G \times G) \times^{(G_{\mathrm{aff}} \times G_{\mathrm{aff}})Z} G_{\mathrm{aff}}$$

by left and right multiplication on G. $\square$

Lemma 3.1.2 *Let G be a connected linear algebraic group, and Z its scheme-theoretic center. Then Z is contained in any Borel subgroup B of G.*

PROOF: Clearly, B contains the unipotent radical $R_u(G)$. Thus, B is the pre-image of a Borel subgroup of $G/R_u(G)$. Also, $Z(G)$ is contained in the pre-image of $Z(G/R_u(G))$. Thus, we may replace G with $G/R_u(G)$, and hence assume that G is reductive. Now choose a maximal torus T of B; then Z is contained in the scheme-theoretic centralizer $C_G(T)$. But the latter equals T, which yields our assertion. $\square$

Next, we note that Theorem 2.2.2 readily implies the following version of a result of Serre (see [Ser01, Proposition 17]):

Corollary 3.1.3 *Let A be an abelian variety, and $\pi : X \to B$ a principal A-bundle over a smooth variety. Then π has a reduction of structure group to a principal A_n-bundle $Y \to B$ for some integer $n > 0$.* $\square$

We also deduce from Theorem 2.2.2 a result due to Serre in characteristic 0 (see [Ser01, Corollaire 1, p. 131]).

Corollary 3.1.4 *Let A and B be abelian varieties, and $\pi : X \to B$ a principal A-bundle. Then X has a structure of abelian variety containing A such that π is the quotient by that subgroup.*

PROOF: By Corollary 3.1.3, we have $X \cong A \times^{A_n} Y$ where Y is a closed subscheme of X stable under A_n, and π restricts to a principal A_n-bundle $Y \to B$. Using a result of Nori (see [Nor83, Remark 2]), we know that such a Y has to be equivariantly isomorphic to $A_n \times^\Gamma Z$, where Γ is a subgroup scheme of A_n and Z is an abelian variety on which Γ acts by translation; then $Z/\Gamma \cong Y/A_n \cong B$. Thus, $X \cong A \times^{A_n} (A_n \times^\Gamma Z) \cong A \times^\Gamma Z$ is the quotient of the abelian variety $A \times Z$ by its subgroup scheme Γ, and π is identified with the quotient morphism $X \to X/A = Z/\Gamma$. $\qquad\square$

The smoothness assumption in Corollary 3.1.3 cannot be suppressed, as shown by the following example of Raynaud (see [Ray70, XIII 3.1]):

Example 3.1.5 Let E, p_0, τ_{p_0} X be as in Example 2.2.3. The projection $E \times \mathbb{P}^1 \to \mathbb{P}^1$ yields a morphism $\pi : X \to C$, where C denotes the nodal curve obtained by identifying 0 and ∞ in $\mathbb{P}^1$. We now show that π is a principal E-bundle for the fppf topology. (See Section 6.1 for details on this notion).

Let $\widetilde{C}$ be the curve obtained as a union of copies of $\mathbb{P}^1$ indexed by $\mathbb{Z}$, such that the point ∞ in each copy is identified with the point 0 in the next copy, as shown in the picture below.

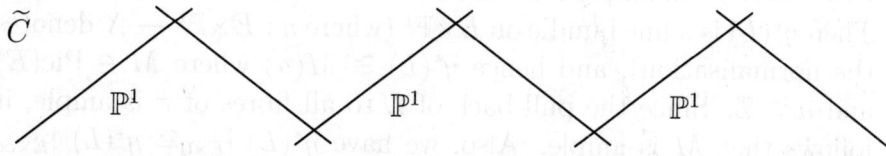

Then $\widetilde{C}$ is a scheme locally of finite type. Let $\tau : \widetilde{C} \to \widetilde{C}$ be the translation that shifts each copy of $\mathbb{P}^1$ to the next copy. Then

$$\widetilde{C}/\tau \cong \mathbb{P}^1/(0 \sim \infty) \cong C$$

and the morphism $\widetilde{C} \to C$ is a étale Galois covering with group $\mathbb{Z}$ (generated by τ). We can now use a similar picture to get a covering $\widetilde{X}$ of X. Here $\widetilde{X}$ consists of copies of $E \times \mathbb{P}^1$ indexed by $\mathbb{Z}$ with successive copies glued to each other as shown in the picture.

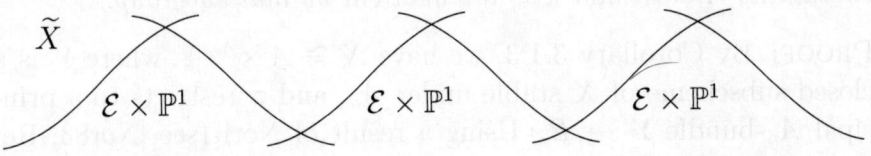

More specifically, each copy of $E \times \mathbb{P}^1$ is mapped to the next one via $(p, z) \mapsto (p + p_0, z^{-1})$. Consider the map $\tau_{p_0} \times \tau : \widetilde{X} \to \widetilde{X}$. We have:

$$\widetilde{X}/(\tau_{p_0} \times \tau) \cong X$$

and the resulting morphism $\widetilde{X} \to X$ is again a $\mathbb{Z}$-cover. The projection $E \times \mathbb{P}^1 \to \mathbb{P}^1$ induces a morphism $\widetilde{\pi} : \widetilde{X} \to \widetilde{C}$ which is a trivial E-bundle. Moreover, the diagram

$$\begin{array}{ccc} \widetilde{X} & \xrightarrow{\widetilde{\pi}} & \widetilde{C} \\ \downarrow & & \downarrow \\ X & \xrightarrow{\pi} & C \end{array}$$

is cartesian, the vertical arrows being the quotients by the $\mathbb{Z}$-action via translations. Thus, π is a principal E-bundle for the fppf topology.

Next, we show that the morphism π is not projective. For otherwise, there is an ample line bundle L on X such that L is π-ample. Then $\eta^*(L)$ is a line bundle on $E \times \mathbb{P}^1$ (where $\eta : E \times \mathbb{P}^1 \to X$ denotes the normalisation), and hence $\eta^*(L) \cong M(n)$ where $M \in \mathrm{Pic}(E)$ and $n \in \mathbb{Z}$. Since the pull-back of L to all fibres of π is ample, it follows that M is ample. Also, we have $\eta^*(L) \mid_{E \times 0} \cong \eta^*(L) \mid_{E \times \infty}$ via τ_{p_0}, i.e., $M \cong \tau_{p_0}^*(M)$. But since M is ample, any $p \in E$ such that $M \cong \tau_p^*(M)$ must have finite order, a contradiction.

Finally, we show that the principal E-bundle π does not satisfy the assertion of Corollary 3.1.3. For otherwise, there exists an E-equivariant morphism $\phi : X \to E/E_n$; then $(\phi, \pi) : X \to E/E_n \times C$ is finite, and hence $\phi^*(M)$ is π-ample for any ample line bundle M on E/E_n. But this contradicts the fact that π is not projective.

By the same arguments, one can show that the restriction $\pi_V : \pi^{-1}(V) \to V$ is not projective whenever V is an open neighborhood of the singular point of C. Moreover, π_V does not satisfy the assertion of Corollary 3.1.3 either.

3.2 "Dual" of Chevalley's theorem and the Rosenlicht decomposition

Recall the statement of Theorem 1.2.1:

Theorem 3.2.1 *Let G be an algebraic group. Then there exists a smallest normal subgroup scheme H of G such that the quotient G/H is affine. Moreover, $\mathcal{O}(H) = k$, and H is the largest subgroup scheme of G satisfying that property; it is in fact a connected algebraic group, contained in the center of G^o. Also, $\mathcal{O}(G/H) = \mathcal{O}(G)$; in particular, the algebra $\mathcal{O}(G)$ is finitely generated.*

PROOF: Let H_1 and H_2 be normal subgroup schemes of G such that G/H_1 and G/H_2 are affine. Consider the morphism

$$p_1 \times p_2 : G/H_1 \cap H_2 \to G/H_1 \times G/H_2,$$

where p_i denotes the quotient morphism $G/H_1 \cap H_2 \to G/H_i$, for $i = 1, 2$. Then $p_1 \times p_2$ is a homomorphism of algebraic groups with trivial scheme-theoretic kernel, and hence is a closed immersion. Thus, $G/H_1 \cap H_2$ is affine. Now choose H_1 minimal among the normal subgroup schemes with affine quotient (such a minimal subgroup scheme exists since G is of finite type). Then we must have $H_1 \cap H_2 = H_1$, i.e., $H_1 \subset H_2$ is the smallest such subgroup. Thus, G has a smallest normal subgroup scheme H such that G/H is affine.

Applying Lemma 2.1.3, we get that G/H^o_{red} is affine; hence $H = H^o_{\mathrm{red}}$. Thus, by the minimality of H, it is connected and reduced. In other words, H is a connected algebraic group; in particular, $H \subset G^o$. Also, since $G^o/Z(G^o)$ is affine by Lemma 2.1.1 and Corollary 2.1.7, we see that H is contained in $Z(G^o)$.

The group H acts on itself by right multiplication, hence it acts also on $\mathcal{O}(H)$. Let K denote the scheme-theoretic kernel for this action. Then, as K is invariant under any automorphism of H, it is a normal subgroup scheme of G. Also, H/K has a faithful rational representation on $\mathcal{O}(H)$; hence H/K is affine by Lemma 2.1.5. So we conclude using Lemma 2.1.1 that G/K is affine. By the choice of H, therefore, we get $H = K$. This implies that H acts trivially on $\mathcal{O}(H)$, so that $\mathcal{O}(H)^H = \mathcal{O}(H)$. However, the only regular functions on H which are invariant under the H-action are constants, i.e., $\mathcal{O}(H)^H = k$. Thus, $\mathcal{O}(H) = k$, i.e., H is anti-affine.

We now show that $\mathcal{O}(G/H) = \mathcal{O}(G)$. For this, note that $\mathcal{O}(G/H) = \mathcal{O}(G)^H$ where H acts on $\mathcal{O}(G)$ via the action induced by the right multiplication on G. The matrix coefficients for the above action give regular functions on H, and hence are constant. Thus we conclude that H acts trivially on $\mathcal{O}(G)$; hence $\mathcal{O}(G/H) = \mathcal{O}(G)^H$ equals $\mathcal{O}(G)$, as required.

It remains to show that H is the largest anti-affine subgroup scheme of G. But for any anti-affine subgroup scheme $K \subset G$, the quotient group scheme $K/K \cap H$ is affine (as a subgroup scheme of G/H) and anti-affine (since $\mathcal{O}(K/K \cap H) \subset \mathcal{O}(K)$), hence a point. So $K \subset H$, as required. $\square$

Remark 3.2.2 The subgroup H in Theorem 1.2.1 gives a morphism $\pi : G \to G/H$, where the quotient G/H being affine is isomorphic to Spec $\mathcal{O}(G/H)$. As also was seen in the above theorem, Spec $\mathcal{O}(G/H) \cong$ Spec $\mathcal{O}(G)$. So π is the canonical morphism $G \to$ Spec $\mathcal{O}(G)$, called the *affinization morphism* (the universal morphism from G to an affine scheme).

Next, we recall the statement of Theorem 1.2.4:

Theorem 3.2.3 *Let G be a connected algebraic group, G_{aff} its largest connected affine normal subgroup, and G_{ant} its largest anti-affine subgroup. Then $G = G_{\mathrm{aff}}G_{\mathrm{ant}}$, and G_{ant} is the smallest subgroup scheme H of G such that $G = G_{\mathrm{aff}}H$, i.e., the restriction to H of the quotient homomorphism $\alpha : G \to G/G_{\mathrm{aff}}$ is surjective. Moreover, the scheme-theoretic intersection $G_{\mathrm{aff}} \cap G_{\mathrm{ant}}$ contains $(G_{\mathrm{ant}})_{\mathrm{aff}}$ as a normal subgroup with finite quotient.*

PROOF: By definition, both G_{aff} and G_{ant} are normal subgroups of G, and thus, so is their product. Hence, on the one hand, considering the group $G/G_{\text{aff}}G_{\text{ant}}$ as a quotient of G/G_{ant} by a normal subgroup, it is affine; on the other hand, considering it as a quotient of G/G_{aff}, it is complete. So $G/G_{\text{aff}}G_{\text{ant}}$ is trivial, *i.e.*, $G = G_{\text{aff}}G_{\text{ant}}$.

Next, let H be a subgroup scheme of G such that $G = G_{\text{aff}}H$, and let $K := H_{\text{red}}^o$. Then

$$G = G_{\text{aff}}K = G_{\text{aff}}K_{\text{aff}}K_{\text{ant}} = G_{\text{aff}}K_{\text{ant}}.$$

By Theorem 1.2.1, K_{ant} is central in G. Moreover, $G/K_{\text{ant}} \cong G_{\text{aff}}/G_{\text{aff}} \cap K_{\text{ant}}$ and the latter quotient is affine. It follows that $G_{\text{ant}} \subset K_{\text{ant}}$, so that $G_{\text{ant}} = K_{\text{ant}} \subset H$.

Finally, consider the connected algebraic group $\bar{G} = G/(G_{\text{ant}})_{\text{aff}}$. Then $\bar{G} = \bar{G}_{\text{aff}}\bar{G}_{\text{ant}}$. Further, the (scheme-theoretic) intersection $\bar{G}_{\text{aff}} \cap \bar{G}_{\text{ant}}$ is both affine and contained in a complete variety, hence finite. Equivalently, the quotient $(G_{\text{aff}} \cap G_{\text{ant}})/(G_{\text{ant}})_{\text{aff}}$ is finite. $\square$

The *Rosenlicht decomposition*

$$G = G_{\text{aff}}G_{\text{ant}} \cong (G_{\text{aff}} \times G_{\text{ant}})/(G_{\text{aff}} \cap G_{\text{ant}})$$

may be combined with known structure results for linear algebraic groups, to obtain insight into the structure of an arbitrary connected algebraic group G. For example, the properties of the radical and unipotent radical of a linear algebraic group immediately yield the following:

Corollary 3.2.4

(i) G has a largest connected solvable normal subgroup, namely, $R(G_{\text{aff}})G_{\text{ant}}$. This is also the smallest normal subgroup scheme H of G such that G/H is semisimple.

(ii) $R_u(G_{\text{aff}})G_{\text{ant}}$ is the smallest normal subgroup scheme H of G such that G/H is reductive.

As another application, we describe the maximal connected solvable subgroups of G:

Corollary 3.2.5 *Any maximal connected solvable subgroup H of G satisfies $H = BG_{\mathrm{ant}}$, where $B := G_{\mathrm{aff}} \cap H$ is a Borel subgroup of G_{aff}. Moreover, H is its own set-theoretic normalizer in G, and $G/H \cong G_{\mathrm{aff}}/B$. In particular, any two maximal connected solvable subgroups are conjugate in G, and these subgroups are parametrized by the flag variety of G_{aff}.*

PROOF: Since G_{ant} is central in G and connected, HG_{ant} is again a connected solvable subgroup of G. By maximality of H, we see that $H \supset G_{\mathrm{ant}}$, and H/G_{ant} is a Borel subgroup of the linear algebraic group $G/G_{\mathrm{ant}} \cong G_{\mathrm{aff}}/G_{\mathrm{aff}} \cap G_{\mathrm{ant}}$. Also, $G_{\mathrm{aff}} \cap G_{\mathrm{ant}}$ is central in G_{aff}, and hence contained in any Borel subgroup of G_{aff} by Lemma 3.1.2. Thus, $H = BG_{\mathrm{ant}}$ for some Borel subgroup B of G_{aff}. Then we have $G_{\mathrm{aff}} \cap H = B(G_{\mathrm{aff}} \cap G_{\mathrm{ant}}) = B$. Therefore, the set-theoretic normalizer $N_G(H)$ satisfies

$$N_G(H) = N_{G_{\mathrm{aff}}}(H)G_{\mathrm{ant}} = N_{G_{\mathrm{aff}}}(G_{\mathrm{aff}} \cap H)G_{\mathrm{ant}} = BG_{\mathrm{ant}} = H,$$

since $N_{G_{\mathrm{aff}}}(B) = B$. Finally, $G/H \cong G_{\mathrm{aff}}/G_{\mathrm{aff}} \cap H = G_{\mathrm{aff}}/B$.   □

Finally, for B as above, we show that G/B is the product of the Albanese variety of G and the flag variety of G_{aff}:

Corollary 3.2.6 *Let H be a subgroup scheme of G. Then the variety G/H is complete if and only if H contains a Borel subgroup of G_{aff}. Moreover,*

$$G/B \cong A \times G_{\mathrm{aff}}/B \tag{3.2}$$

for any such Borel subgroup B, where A denotes the abelian variety G/G_{aff}.

PROOF: If G/H is complete, then it contains a fixed point of B by Borel's fixed point theorem, *i.e.*, H contains a conjugate of B. For the converse, it suffices to prove the isomorphism (3.2): indeed, this isomorphism implies that G/B is complete, and hence so is its image G/H under the natural morphism $G/H \to G/B$. Now

$$G_{\mathrm{aff}}/B = G_{\mathrm{aff}}/G_{\mathrm{aff}} \cap BG_{\mathrm{ant}} \cong G/BG_{\mathrm{ant}}$$

where the equality follows from Corollary 3.2.5, and the isomorphism, from the Rosenlicht decomposition. Thus,

$$G/B \cong G \times^{G_{\mathrm{aff}}} G_{\mathrm{aff}}/B \cong G \times^{G_{\mathrm{aff}}} G/BG_{\mathrm{ant}} \cong G/G_{\mathrm{aff}} \times G/BG_{\mathrm{ant}}$$
$$\cong A \times G_{\mathrm{aff}}/B,$$

where the third isomorphism follows from the isomorphism

$$G \times^{H} X \cong G/H \times X, \quad (g, x)H \mapsto (gH, gx),$$

for any subgroup scheme H of G and for any G-scheme X. $\qquad\square$

This corollary is the starting point for the investigation of the Chow ring of G; in particular, of its Picard group (see [Bri1, Sections 2.2, 2.3]). The product structure of G/B will be generalized to all complete homogeneous varieties in the next chapter.

Chapter 4

Structure and automorphisms of complete homogeneous varieties

In this chapter we shall prove the structure Theorem 1.3.1 for complete homogeneous varieties and their automorphism group schemes, along the lines of the proof in [Bri10a]. We begin by decomposing any complete homogeneous variety as the product of an abelian variety and a homogeneous variety under an affine group (Theorem 4.1.1). Then we turn to automorphisms: we show, quite generally, that the connected automorphism group scheme of a product of complete varieties is the product of their connected automorphism group schemes (Corollary 4.2.7). We also prove that the connected automorphism group scheme of an abelian variety is just its group of translations (Proposition 4.3.2). Putting these results together completes the proof of Theorem 1.3.1.

We conclude this chapter with some considerations on the automorphism group schemes of complete varieties that are homogeneous under an affine group. We show by an example that these group schemes may be non-reduced for varieties of unseparated flags (Proposition 4.3.4).

4.1   Structure

The following result is the first step in the proof of Theorem 1.3.1:

Theorem 4.1.1 *Let X be a complete variety, homogeneous under a scheme-theoretic faithful action of a connected algebraic group G. Then there exist canonical isomorphisms*

$$G \cong A \times G_{\mathrm{aff}} \tag{4.1}$$

of algebraic groups and

$$X \cong A \times Y \tag{4.2}$$

of G-varieties, where A is an abelian variety, G_{aff} a semisimple group of adjoint type, and Y a complete homogeneous G_{aff}-variety; moreover, G acts on X via the action of A on itself by translation, and the given action of G_{aff} on Y.

PROOF: We first show (4.1). Let B be a Borel subgroup of G_{aff}. Since X is complete, B fixes some point $x \in X$. The subgroup scheme $G_{\mathrm{aff}} \cap G_{\mathrm{ant}}$ is central in G, and hence contained in B by Lemma 3.1.2; thus, $G_{\mathrm{aff}} \cap G_{\mathrm{ant}}$ fixes x. *However, being a normal subgroup scheme of G and X being G-homogeneous, $G_{\mathrm{aff}} \cap G_{\mathrm{ant}}$ also fixes every point of X. The action of G being faithful, this can happen only if the subgroup scheme is trivial* (†). Thus, along with the Rosenlicht decomposition (Theorem 1.2.4), we see that $G \cong G_{\mathrm{ant}} \times G_{\mathrm{aff}}$. Moreover, $(G_{\mathrm{ant}})_{\mathrm{aff}}$ is a subgroup of $G_{\mathrm{aff}} \cap G_{\mathrm{ant}}$ and hence trivial; thus G_{ant} is an abelian variety, which we denote by A. This establishes (4.1).

The radical $R(G_{\mathrm{aff}})$ is contained in the Borel subgroup B so it fixes x. By an argument as in (†) above, applied to $R(G_{\mathrm{aff}})$, we conclude that $R(G_{\mathrm{aff}})$ is trivial. Thus, G_{aff} is semi-simple. Next, consider the scheme-theoretic center $Z(G_{\mathrm{aff}})$. Since $Z(G_{\mathrm{aff}})$ is contained in B (by Lemma 3.1.2 again), it fixes x; by the above argument, $Z(G_{\mathrm{aff}})$ must be trivial, *i.e.*, G_{aff} is of adjoint type.

We now show (4.2). Let H denote the isotropy subgroup scheme G_x. By an argument as in (†) above applied to the scheme-theoretic intersection $A \cap H = A_x$, we conclude that $A \cap H$ is trivial. Hence the projection $p_2 : G \to G_{\mathrm{aff}}$ when restricted to H is (scheme-theoretic) faithful. Now note that H contains B, and is affine by Proposition 2.1.6. Thus, $B \subset H_{\mathrm{red}}^o \subset G_{\mathrm{aff}}$; in particular, H_{red}^o is a parabolic subgroup of G_{aff}, denoted as P. Since P is its own set-theoretic normalizer in G_{aff}, we have $p_2(H)_{\mathrm{red}} = P$; as $p_2 \mid_{H_{\mathrm{red}}}$ is faithful, it follows that $H_{\mathrm{red}} = P$.

In characteristic 0, we even have $H = P$ and hence

$$G/H \cong G/P \cong A \times G_{\mathrm{aff}}/P,$$

as required. When $\mathrm{char}(k) = p > 0$, we may have $H \neq P$ (see Example 1.3.2) but we shall show that $H \subset G_{\mathrm{aff}}$ and hence $G/H \cong A \times G_{\mathrm{aff}}/H$.

Let $P^- \subset G_{\mathrm{aff}}$ be a parabolic subgroup opposite to P, *i.e.*, P^- is a parabolic subgroup such that $P^- \cap P$ is a Levi subgroup of both P^- and P. Then the multiplication of G yields an open immersion

$$R_u(P^-) \times P \to G_{\mathrm{aff}}.$$

In particular, $R_u(P^-)P$ is an open neighbourhood of P in G_{aff}. Hence $A \times R_u(P^-)P$ is an open neighbourhood of H_{red} in $A \times G_{\mathrm{aff}} \cong G$. This is also an open neighbourhood of H in G, isomorphic to $A \times R_u(P^-) \times P$. Since $H_{\mathrm{red}} = P$, this yields the decomposition $H = \Gamma P$, where $\Gamma := (A \times R_u(P^-)) \cap H$ is a finite subgroup scheme of G. Now choose a maximal torus T of $P \cap P^-$; then T acts on $A \times R_u(P^-)$ by conjugation, and the categorical quotient is the projection

$$p_2 : R_u(P^-) \times A \to A$$

with image the T-fixed point subscheme. Thus, for the closed T-stable subscheme $\Gamma \subset A \times R_u(P^-)$, the quotient is the restriction of p_2 to Γ with image $\Gamma^T = A \cap \Gamma$. But the latter is a closed subscheme of $A \cap H$, which is just the origin of A (viewed as a reduced subscheme). Hence Γ is contained in $R_u(P^-)$, and $H = \Gamma P$ is contained in G_{aff} as required.

As the isotropy subgroup scheme of any point of X is contained in G_{aff}, the uniqueness of the decomposition (4.2) follows from the uniqueness of the decomposition (4.1). $\qquad\square$

4.2   Blanchard's lemma for group schemes

The following proposition is a version of a result of Blanchard about holomorphic transformation groups (see [Bla56, §I.1] and also [Akh95, §2.4]) in the setting of actions of group schemes.

Proposition 4.2.1 *Let* $f : X \to Y$ *be a proper morphism of schemes such that* $f_*(\mathcal{O}_X) = \mathcal{O}_Y$. *Let* G *be a connected group scheme acting on* X. *Then there exists a unique* G-*action on* Y *such that* f *is* G-*equivariant.*

PROOF: As schemes are assumed to be finite type over the algebraically closed field k, we shall consider them as ringed spaces; specifically, the scheme X shall be considered as the set $X(k)$ equipped with the Zariski topology and with the structure sheaf $\mathcal{O}_X$ ([DG70, §I.3.6]).

Step 1: the group $G(k)$ **acts on** $Y(k)$ **(viewed as an abstract set) such that** f **is equivariant.** We first show that $G(k)$ permutes the fibres of f. The conditions that f is proper and $f_*(\mathcal{O}_X) = \mathcal{O}_Y$ imply that the set-theoretic fibre of f at any $y \in Y(k)$ is non-empty, complete, and connected (if f is projective, the latter property follows from [Har77, Corollary III.11.3]; the general case reduces to that one by Chow's lemma, see [Har77, Exercise II.4.10]). Let F_y denote that fibre considered as a reduced subscheme of X. The map $\phi : G_{\mathrm{red}} \times F_y \to Y$ given by $(g, x) \mapsto f(g.x)$ sends $\{e_G\} \times F_y$ to $\{y\}$. As F_y is complete and G_{red} is a variety, we apply the rigidity lemma ([Mil86, Theorem 2.1]) to ϕ in order to conclude that $\phi(g \times F_y)$ is a single point. Thus, $g.F_y$ is contained in the fibre over a single point, for each $g \in G_{\mathrm{red}}(k)$; denote this point as $g.y$. Then $g^{-1}.F_{g.y} \subset F_y$ and hence $g.F_y = F_{g.y}$. It is straightforward to check that the map $(g, y) \mapsto g.y$ indeed gives an action of the abstract group $G(k) = G_{\mathrm{red}}(k)$ on the set $Y(k)$.

The above claim gives the commutativity of the following diagram (as sets)

$$\begin{array}{ccc} G(k) \times X(k) & \xrightarrow{\ \alpha\ } & X(k) \\ {\scriptstyle \mathrm{id} \times f} \downarrow & & \downarrow {\scriptstyle f} \\ G(k) \times Y(k) & \xrightarrow{\ \beta\ } & Y(k) \end{array}$$

where α, β are the action maps of $G(k)$ on $X(k)$ and $Y(k)$ respectively.

Step 2: The map β is continuous. It suffices to check that $\beta^{-1}(Z)$ is closed for any closed subset Z of $Y(k)$. By the commutativity of the above diagram, we have

$$(\text{id} \times f)^{-1}\beta^{-1}(Z) = \alpha^{-1}f^{-1}(Z).$$

As α, f are morphisms of schemes, in particular continuous, the subset $(\text{id} \times f)^{-1}\beta^{-1}(Z)$ of $G(k) \times X(k)$ is closed. The map $\text{id} \times f$ being surjective and proper, we therefore get that $\beta^{-1}(Z)$ is closed.

We have shown, so far, that the map β is an action of the topological group $G(k)$ on the topological space $Y(k)$.

Step 3: Construction of the morphism of sheaves $\beta^{\#}$: $\mathcal{O}_Y \to \beta_*(\mathcal{O}_{G \times Y})$. Let V be an open subset of $Y(k)$. We have to define a map

$$\beta^{\#}(V) : \mathcal{O}_Y(V) \to \mathcal{O}_{G \times Y}(\beta^{-1}(V)).$$

By assumption the left hand side is $(f_*\mathcal{O}_X)(V) = \mathcal{O}_X(f^{-1}(V))$ while the right hand side is $(\text{id} \times f)_*\mathcal{O}_{G \times X}(\beta^{-1}(V))$, which by the commutativity of the above diagram is the same as $\mathcal{O}_{G \times X}(\alpha^{-1}f^{-1}(V))$. Hence, we define $\beta^{\#}(V)$ to be $\alpha^{\#}(f^{-1}(V))$. It is obvious that $\beta^{\#}$ is indeed a morphism of sheaves.

All the axioms to be satisfied by β for it to define an action of the group scheme G on Y can also be verified similarly. $\qquad\square$

Remark 4.2.2 In the above statement, the assumption that $f_*(\mathcal{O}_X) = \mathcal{O}_Y$ cannot be omitted. For example, if E is an elliptic curve and $f : E \to \mathbb{P}^1$ the quotient by the involution $(-1)_E$, then the action of E on itself by translation does not descend to an action on $\mathbb{P}^1$.

We now apply this result to automorphisms group schemes; these schemes are defined as follows. For a scheme X we have a contravariant functor

$$S \mapsto \text{Aut}_S(X \times S)$$

denoted as $\underline{\mathrm{Aut}}(X)$, from the category of schemes to the category of groups; here $\mathrm{Aut}_S(X \times S)$ denotes the group of automorphisms of $X \times S$ viewed as a scheme over S. If X is proper (in particular, if X is a complete variety) then $\underline{\mathrm{Aut}}(X)$ is represented by a group scheme, $\mathrm{Aut}(X)$, locally of finite type over k; in particular, the neutral component $\mathrm{Aut}^\circ(X)$ is of finite type (see [MO67, Theorem 3.7]). The Lie algebra of $\mathrm{Aut}(X)$, or equivalently of $\mathrm{Aut}^\circ(X)$, is identified with the Lie algebra of derivations of $\mathcal{O}_X$; in other words, of global vector fields on X. If X is smooth, then

$$\mathrm{Lie}\,\mathrm{Aut}(X) \cong \mathrm{H}^0(X, T_X)$$

(the Lie algebra of global sections of the tangent sheaf); see [DG70, §II.4.2] for details.

Example 4.2.3 Let C be a smooth projective curve, and g its genus. If $g = 0$, *i.e.*, $X \cong \mathbb{P}^1$, then $\mathrm{Aut}(C) \cong \mathrm{PGL}_2$. If $g = 1$, *i.e.*, C is an elliptic curve, then $\mathrm{Aut}(C)$ is the semi-direct product of C (acting on itself by translations) and of $\mathrm{Aut}_{\mathrm{gp}}(C)$ (the group of automorphisms of X as an algebraic group), where $\mathrm{Aut}_{\mathrm{gp}}(C)$ is viewed as a constant group scheme; this result holds in fact for all abelian varieties, see Proposition 4.3.2 below. Moreover, the group $\mathrm{Aut}_{\mathrm{gp}}(C)$ is finite by [Har77, Corollary IV.4.7]. Finally, if $g \geq 2$ then $\mathrm{Aut}(C)$ is finite and reduced, since $\mathrm{H}^0(C, T_C) = 0$ by the Riemann-Roch theorem.

Example 4.2.4 The automorphism group schemes of smooth projective surfaces yield examples of non-trivial anti-affine groups. Specifically, let $p : E \to C$ be a vector bundle of rank 2 over an elliptic curve, and $\pi : S = \mathbb{P}(E) \to C$ the associated ruled surface.

Assume that $E \cong L \oplus \mathcal{O}_C$, where L is a line bundle of degree 0 and $\mathcal{O}_C$ denotes the trivial line bundle. Then one can show that $\mathrm{Aut}^\circ(S)$ is a commutative algebraic group, and sits in an exact sequence

$$0 \to \mathbb{G}_m \to \mathrm{Aut}^\circ(S) \xrightarrow{\pi_*} C \to 0.$$

Moreover, $\mathrm{Aut}^\circ(S)$ acts on S with 3 orbits: the two sections C_1, C_2 of π associated to the sub-bundles L, $\mathcal{O}_C$ of E, and the open subset $S \setminus (C_1 \cup C_2)$, isomorphic to $\text{Aut}^\circ(S)$. In particular, $\text{Aut}^\circ(S)$ is isomorphic to the principal $\mathbb{G}_m$-bundle associated to L; it is anti-affine if and only if L has infinite order in $\text{Pic}^\circ(C) \cong C$.

Next, assume that E is indecomposable and sits in an exact sequence $0 \to \mathcal{O}_C \to E \to \mathcal{O}_C \to 0$ (such a bundle exists since $\text{H}^1(C, \mathcal{O}_C) \cong k$). Then one can show that $\text{Aut}^\circ(S)$ is again a commutative algebraic group, and sits in an exact sequence

$$0 \to \mathbb{G}_a \to \text{Aut}^\circ(S) \xrightarrow{\pi_*} C \to 0.$$

Moreover, $\text{Aut}^\circ(S)$ acts on S with 2 orbits: the section C_1 associated to the sub-bundle $\mathcal{O}_C$ of E, and the open subset $S \setminus C_1$ isomorphic to $\text{Aut}^\circ(S)$. It follows that $\text{Aut}^\circ(S)$ is the universal vector extension of C; it is anti-affine if and only if $\text{char}(k) = 0$.

These results follow from [Mar71, Theorem 3], which describes the automorphism group schemes of all ruled surfaces; refer to Example 1.2.3 for the claim about the anti-affineness.

Remark 4.2.5 The group scheme $\text{Aut}(X)$ is generally non-reduced in positive characteristics. This already happens for certain smooth complete surfaces, see [MO67, Example 6]; we shall see in Proposition 4.3.4 that this also happens for certain complete homogeneous varieties. Also, for a non-proper scheme X, the group functor $\underline{\text{Aut}}(X)$ is generally not representable, already for the affine space $\mathbb{A}^n$; then $\text{Aut}(X)$ can be given the structure of an ind-algebraic group, see [Kam79, Theorem 1.1].

The category of schemes is a fully faithful subcategory of the category of functors from schemes to sets, such that group schemes correspond, under this identification, to functors from schemes to groups (see [DG70, §II.1.1] for more details); further, an isomorphism of schemes corresponds to an isomorphism of the associated functors. In particular, when X and Y are proper schemes, $\text{Aut}(X) \cong \text{Aut}(Y)$ if and only if $\underline{\text{Aut}}(X) \cong \underline{\text{Aut}}(Y)$. We use this approach to prove the following:

Corollary 4.2.6 *Let* $f : X \to Y$ *be a proper morphism of schemes such that* $f_*(\mathcal{O}_X) = \mathcal{O}_Y$. *Then* f *induces a homomorphism of group schemes*

$$f_* : \text{Aut}^\circ(X) \to \text{Aut}^\circ(Y).$$

PROOF: Let $G := \text{Aut}^\circ(X)$. By the above proposition, we have a G-action on Y and hence an automorphism of $Y \times G$ as a scheme over G, namely, the morphism $(y, g) \mapsto (g \cdot y, g)$. This yields in turn a morphism of schemes $f_* : G \to \text{Aut}(Y)$, that sends e_G to the identity. Since G is connected, it follows that the image of f_* is contained in $\text{Aut}^\circ(Y)$. Also, since f_* corresponds to the G-action on Y, it is a homomorphism of group functors, and hence of group schemes. □

Note that when X and Y are smooth, the morphism of Lie algebras induced by f_* is just the map $\text{H}^0(X, T_X) \to \text{H}^0(Y, T_Y)$ induced by the differential $df : T_X \to f^*(T_Y)$ and by the isomorphisms

$$\text{H}^0(X, f^*T_Y) = \text{H}^0(Y, f_*(f^*T_Y)) \cong \text{H}^0(Y, T_Y \otimes f_*\mathcal{O}_X) \cong \text{H}^0(Y, T_Y).$$

We may now describe the connected automorphism group scheme of a product of complete varieties:

Corollary 4.2.7 *For any complete varieties X and Y, we have a canonical isomorphism*

$$\text{Aut}^\circ(X \times Y) \cong \text{Aut}^\circ(X) \times \text{Aut}^\circ(Y).$$

PROOF: Denote by $p_1 : X \times Y \to X$ and $p_2 : X \times Y \to Y$ the projections. Then $(p_1)_*(\mathcal{O}_{X \times Y}) = \mathcal{O}_X \otimes \mathcal{O}(Y) = \mathcal{O}_X$ as Y is complete; similarly, $(p_2)_*(\mathcal{O}_{X \times Y}) = \mathcal{O}_Y$. So, applying the above corollary, we get a homomorphism of group schemes

$$(p_1)_* \times (p_2)_* : \text{Aut}^\circ(X \times Y) \to \text{Aut}^\circ(X) \times \text{Aut}^\circ(Y).$$

This has an inverse given by the natural homomorphism that sends $(\phi, \psi) \in \text{Aut}^\circ(X) \times \text{Aut}^\circ(Y)$ to the automorphism of $X \times Y$ given by $(x, y) \mapsto (\phi(x), \psi(y))$, thus proving the claim. □

Remark 4.2.8 The above corollary does not extend to (say) proper schemes. Consider for example a smooth complete variety X, and let $Y = \text{Spec}(k[t]/(t^2))$. Then $\text{Aut}(Y) = \mathbb{G}_m$, and $\text{Aut}^\circ(X \times Y)$ contains the vector group $\text{H}^0(X, T_X)$ that fixes pointwise $X \cong X \times \text{Spec}(k[t]/(t))$. Thus, $\text{H}^0(X, T_X)$ intersects trivially with the subgroup scheme $\text{Aut}^\circ(X) \times \text{Aut}^\circ(Y)$ of $\text{Aut}^\circ(X \times Y)$.

4.3 Automorphisms of complete homogeneous varieties

We begin this section by describing the automorphism group schemes of abelian varieties; the result is well-known and may be found e.g. in [MO67, Example 1].

Let A be an abelian variety, and denote as $\mathrm{Aut}_{\mathrm{gp}}(A)$ the automorphism group of the algebraic group A; then $\mathrm{Aut}_{\mathrm{gp}}(A)$ is a subgroup of $\mathrm{GL}_N(\mathbb{Z})$ for some N (as follows from [Mil86, Theorem 12.5]). Also, recall that each automorphism of the variety A is of the form $\tau_a \circ f$ for a unique point $a \in A$ and a unique $f \in \mathrm{Aut}_{\mathrm{gp}}(A)$ (see [Mil86, Corollary 2.2]). Since $f^{-1} \circ \tau_a \circ f = \tau_{f(a)}$ for all $a \in A$ and $f \in \mathrm{Aut}_{\mathrm{gp}}(A)$, we see that

$$\mathrm{Aut}(A) \cong A \rtimes \mathrm{Aut}_{\mathrm{gp}}(A)$$

as abstract groups. We shall obtain a scheme-theoretic version of that isomorphism. For this, we need a generalization of the classical rigidity lemma, due to C. and F. Sancho de Salas (see [SS09, Theorem 1.7]) and which will also be used in the next chapter.

Proposition 4.3.1 *Let X, Y and Z be schemes. Assume that X is anti-affine, Y is connected, and Z is separated. Let $f : X \times Y \to Z$ be a morphism such that there exist $y_0 \in Y$ and $z_0 \in Z$ satisfying $f(X \times y_0) = z_0$. Then there exists a unique morphism $g : Y \to Z$ such that $f = g \circ p_2$, where $p_2 : X \times Y \to Y$ denotes the projection.*

PROOF: Choose $x_0 \in X$ and let $g := f(x_0, -) : Y \to Z$. We claim that $f = g \circ p_2$.

Step 1: the claim holds if Z is affine. Then f corresponds to a homomorphism of algebras

$$f^{\#} : \mathcal{O}(Z) \to \mathcal{O}(X \times Y).$$

Moreover, $\mathcal{O}(X \times Y) \cong \mathcal{O}(X) \otimes \mathcal{O}(Y) \cong \mathcal{O}(Y)$ via $p_2^{\#}$ and hence $f = h \circ p_2$ for a unique morphism $h : Y \to Z$. Then $h = f(x_0, -) = g$.

Step 2: it suffices to show that f is constant on all set-theoretic fibres of p_2. Let V be an open affine subset of Z, and $U := g^{-1}(V)$. Then U is open in Y, and $f^{-1}(V) = X \times U$ as sets, assuming that f is constant on set-theoretic fibres of p_2. It follows that f restricts to a morphism $f_U : X \times U \to V$. By Step 1, we have $f_U = g_U \circ p_2$ with an obvious notation. Since Y is covered by open subsets of the form $g^{-1}(V)$ where V is as above, this yields the statement.

Step 3: it suffices to show the claim when Y is irreducible. Assume that the claim holds for all irreducible components of Y, and let Y_0 be an irreducible component containing y_0. If $Y = Y_0$, then there is nothing to prove; otherwise, we may choose another irreducible component Y_1 that meets Y_0 (since Y is connected). Then f is constant on all set-theoretic fibres of p_2 over $Y_0 \cap Y_1$, and hence over Y_1 by the assumption. Arguing by induction, it follows that f is constant on all set-theoretic fibres of p_2. We conclude by Step 2.

Step 4: completion of the proof. We may assume that Y is irreducible. Let

$$W := \{(x, y) \in X \times Y \mid f(x, y) = g(y)\}.$$

Then W is the pre-image of the diagonal in $Z \times Z$ under the morphism $Y \to Z \times Z$, $(x, y) \mapsto (f(x, y), g(y))$. Hence W is a closed subscheme of $X \times Y$. Also, W contains $X \times y_0$ by assumption. Consider the local ring $\mathcal{O}_{Y,y_0}$ and its maximal ideal $\mathfrak{m}_{y_0}$. For any integer $n \geq 1$, denote by

$$Y_n := \mathrm{Spec}(\mathcal{O}_{Y,y_0}/\mathfrak{m}_{y_0}^n)$$

the n-th infinitesimal neighborhood of y_0 in Y (in particular, Y_1 is the reduced point y_0); similarly define Z_n, $n \geq 1$. Then f restricts to a morphism

$$f_n : X \times Y_n \to Z_n.$$

But Z_n is affine, and hence $f_n = g_n \circ p_2$ for some morphism $g_n : Y_n \to Z_n$. Then $g_n(y) = f_n(x_0, y) = f(x_0, y) = g(y)$, so that $f = g \circ p_2$ on $X \times Y_n$; in other words, W contains $X \times Y_n$ for all n. Now the union of all the Y_n is dense in a neighborhood of y_0 in Y, since $\bigcap_{n \geq 1} \mathfrak{m}_y^n = 0$. Thus, W contains a neighborhood of $X \times y_0$. Since W is closed and Y is irreducible, this completes the proof of the claim, and hence of the proposition. $\square$

Proposition 4.3.2 *Let A be an abelian variety. Then we have an isomorphism of group schemes*

$$\mathrm{Aut}(A) = A \rtimes \mathrm{Aut}_{\mathrm{gp}}(A)$$

where A is viewed as the group scheme of translations, and $\mathrm{Aut}_{\mathrm{gp}}(A)$ as a constant group scheme. In particular, $\mathrm{Aut}^{\circ}(A) = A$.

PROOF: Let S be a scheme, $f : A \times S \to A \times S$ an S-automorphism, and $\varphi := p_1 \circ f$ where $p_1 : A \times S \to A$ denotes the projection. Then we have $f(a, s) = (\varphi(a, s), s)$. Choose a point $s_0 \in S$ and denote by T the connected component of s_0 in S. Then the morphism

$$\psi : A \times T \to A, \quad (a, s) \mapsto \varphi(a, s) - \varphi(a, s_0)$$

satisfies the assumptions of the above rigidity result. Thus, there exists a morphism $g : T \to A$ such that $\varphi(a, s) = g(s) + \varphi(a, s_0)$. Moreover, the morphism $a \mapsto \varphi(a, s_0)$ is an automorphism of the variety A, and hence $a \mapsto \varphi(a, s_0) - \varphi(0, s_0)$ yields an automorphism of the group A. Replacing g with $g + \varphi(0, s_0)$, we thus have

$$\varphi(a, s) = g(s) + h(a)$$

where $g \in A(T)$ and $h \in \mathrm{Aut}_{\mathrm{gp}}(A)$. Then $g = \varphi(0, s)$ and hence g, h are uniquely determined by φ. $\square$

We may now complete the proof of Theorem 1.3.1.

Proposition 4.3.3 *With notation and assumptions as in Theorem 4.1.1, we have a canonical isomorphism of group schemes*

$$\mathrm{Aut}^{\circ}(X) \cong A \times \mathrm{Aut}^{\circ}(Y).$$

PROOF: This follows by combining Theorem 4.1.1, Corollary 4.2.7 and Proposition 4.3.2. □

This reduces the description of the connected automorphism group scheme of a complete homogeneous variety X to the case that the acting group G is affine. Then G is semisimple of adjoint type by Theorem 4.1.1; moreover, by Borel's fixed point theorem, $X \cong G/H$ where H_{red} is a parabolic subgroup of G.

If H is reduced (for example, in characteristic 0), then $\mathrm{Aut}(X)$ is an affine algebraic group; moreover, $\mathrm{Aut}^\circ(X)$ is semisimple of adjoint type, and the natural homomorphism $G \to \mathrm{Aut}^\circ(X)$ is an isomorphism when X is "non-exceptional", the exceptional cases being explicitly described (see [Dem77] for these results). An example of an exceptional variety is any odd-dimensional projective space $\mathbb{P}^{2n-1}$ equipped with the action of the projective symplectic group PSp_{2n}, where $n \geq 2$.

For an arbitrary isotropy subgroup scheme H, the connected algebraic group $\mathrm{Aut}^\circ(X)_{\mathrm{red}}$ is still semisimple of adjoint type, by Theorem 4.1.1 again. But $\mathrm{Aut}^\circ(X)$ may be non-reduced, as shown by the following example.

Assume that $\mathrm{char}(k) = p \geq 3$. Let $G = \mathrm{PSp}_{2n}$ and let P_1 be the stabilizer of a line V_1 in the standard representation k^{2n} of G; then P_1 is a maximal parabolic subgroup of G, and $G/P_1 \cong \mathbb{P}^{2n-1}$. Let P_2 denote the stabilizer of a 2-dimensional subspace V_2 of k^{2n}, such that V_2 contains V_1 and is isotropic relative to the symplectic form defining PSp_{2n}. Then P_2 is also a maximal parabolic subgroup of G, and G/P_2 parametrizes the isotropic planes in k^{2n}. Denote by $G_{(1)}$ the kernel of the Frobenius homomorphism $F : G \to G$, and consider $X := G/H$ where $H := P_1 \cap G_{(1)}P_2$. Then $H_{\mathrm{red}} = P_1 \cap P_2$ contains a Borel subgroup B of G, and hence X is complete.

We may view X as the variety of pairs (E_1, E_2) of subspaces of k^{2n} such that E_1 is a line, E_2 is an isotropic plane, and E_2 contains $F(E_1)$; here F denotes the Frobenius endomorphism of $\mathbb{P}^{2n-1}$ that raises homogeneous coordinates to their p-th powers. Then G acts on X via $g \cdot (E_1, E_2) = (g(E_1), F(g)(E_2))$.

Proposition 4.3.4 *With the above notation, the natural homo-morphism $G \to \mathrm{Aut}^\circ(X)$ yields an isomorphism $G \cong \mathrm{Aut}^\circ(X)_{\mathrm{red}}$, but the induced homomorphism of Lie algebras $\mathrm{Lie}(G) \to \mathrm{H}^0(X, T_X)$ is not surjective. In particular, $\mathrm{Aut}^\circ(X)$ is not reduced.*

PROOF: Denote by

$$p_1 : X \to G/P_1, \quad p_2 : X \to G/G_{(1)}P_2$$

the natural morphisms. Then p_1, p_2 are equivariant, and the product $p_1 \times p_2$ is a closed immersion. Moreover, p_1 is a locally trivial fibration for the Zariski topology, with fibre $P_1/(P_1 \cap G_{(1)}P_2)$. This fibre is isomorphic to $\mathbb{P}^{2n-3}$ where P_1 acts through the natural action of its quotient PSp_{2n-2} twisted by the Frobenius F. In particular, $(p_1)_*(\mathcal{O}_X) = \mathcal{O}_{G/P_1}$; by Proposition 4.2.1, this yields a homomorphism of group schemes

$$(p_1)_* : \mathrm{Aut}^\circ(X) \to \mathrm{Aut}^\circ(G/P_1) = \mathrm{PGL}_{2n}. \tag{4.3}$$

The induced homomorphism of Lie algebras is the differential

$$dp_1 : \mathrm{H}^0(X, T_X) \to \mathrm{H}^0(G/P_1, T_{G/P_1}) = \mathrm{Lie}(\mathrm{PGL}_{2n}) \tag{4.4}$$

arising from the exact sequence of tangent sheaves

$$0 \to T_{p_1} \to T_X \to p_1^*(T_{G/P_1}) \to 0 \tag{4.5}$$

where T_{p_1} denotes the relative tangent bundle.

We claim that the exact sequence (4.5) is split. To see this, consider the subsheaf $\mathcal{F}$ of T_X generated by the Lie algebra $\mathrm{Lie}(G)$, acting on X by global vector fields. Then the restriction $\mathcal{F} \to p_1^*(T_{G/P_1})$ is surjective, since T_{G/P_1} is generated by $\mathrm{Lie}(G)$. Moreover, $p_2 : X \to G/G_{(1)}P_2$ being $G_{(1)}$-invariant, its differential is $\mathrm{Lie}(G)$-invariant; thus, $\mathcal{F}$ is contained in the relative tangent sheaf T_{p_2}. But $T_{p_2} \cap T_{p_1} = 0$ since $p_1 \times p_2$ is an immersion; thus, the restriction $\mathcal{F} \to p_1^*(T_{G/P_1})$ is injective as well. So $\mathcal{F}$ yields the required splitting.

By the claim, the map dp_1 is surjective. We now claim that it is injective, i.e., $\mathrm{H}^0(X, T_{p_1}) = 0$. Indeed, $\mathrm{H}^0(X, T_{p_1}) =$

$H^0(G/P_1, (p_1)_*(T_{p_1}))$ and $(p_1)_*(T_{p_1})$ is the homogeneous vector bundle on G/P_1 associated to the P_1-module $H^0(P_1/P_1 \cap G_{(1)}P_2, T_{p_1})$. This P_1-module is isomorphic to $\mathrm{Lie}(\mathrm{PGL}_{2n-2})$ on which P_1 acts via conjugation by its quotient PSp_{2n-2}, twisted by F. Moreover, we have isomorphisms of PGL_{2n}-modules

$$\mathrm{Lie}(\mathrm{PGL}_{2n}) \cong \mathbb{M}_{2n-2}/k \ \mathrm{id} \cong ((k^{2n-2})^* \otimes k^{2n-2})/k$$

and $(k^{2n-2})^* \cong k^{2n-2}$ as PSp_{2n-2}-modules, via the symplectic form. It follows that

$$\mathrm{Lie}(\mathrm{PGL}_{2n}) \cong (k^{2n-2} \otimes k^{2n-2})/k \cong S^2(k^{2n-2}) \oplus (\Lambda^2(k^{2n-2})/k),$$

where S^2 (resp. Λ^2) denotes the symmetric (resp. alternating) square, so that the PSp_{2n-2}-module $\Lambda^2(k^{2n-2})$ contains the trivial module k. This decomposes the P_1-module $H^0(P_1/P_1 \cap G_{(1)}P_2, T_{p_1})$ into the direct sum of two simple modules, $H^0(P_1/B, L_\chi)$ and $H^0(P_1/B, L_{\chi'})$, where χ, χ' are characters of B, dominant for P_1, and L_χ, $L_{\chi'}$ denote the associated homogeneous line bundles. So

$$H^0(X, T_{p_1}) = H^0(G/B, L_\chi) \oplus H^0(G/B, L_{\chi'}).$$

But one checks that neither χ nor χ' are dominant for G, which yields the second claim.

Combining both claims, we see that dp_1 is an isomorphism, and hence the scheme-theoretic kernel of $(p_1)_*$ is trivial; it follows that $\mathrm{PSp}_{2n} \subset \mathrm{Aut}^\circ(X)_{\mathrm{red}} \subset \mathrm{PGL}_{2n}$. Since PSp_{2n} is a maximal subgroup of PGL_{2n}, the proof will be completed if we show that $\mathrm{Aut}^\circ(X)_{\mathrm{red}} \neq \mathrm{PGL}_{2n}$. But otherwise, PGL_{2n} acts on X, compatibly with its action on $\mathbb{P}^{2n-1}$ in view of Theorem 4.2.1. Let P denote the maximal parabolic subgroup of PGL_{2n} that stabilizes V_1, so that $\mathbb{P}^{2n-1} \cong \mathrm{PGL}_{2n}/P$. Then $X \cong \mathrm{PGL}_{2n} \times^P Y$, where Y is a P-variety (the fibre of p_1 at V_1). Then Y must be homogeneous under the action of P, and hence under an action of PGL_{2n-1} (the largest semisimple quotient of adjoint type of P). But $Y \cong \mathbb{P}^{2n-3}$ which yields a contradiction, since PGL_{2n-1} contains no parabolic subgroup of codimension $2n-3$. $\qquad\square$

Chapter 5

Anti-affine groups

Recall that a scheme X is said to be anti-affine if $\mathcal{O}(X) = k$. In this chapter, we obtain characterizations of anti-affine group schemes, and we describe their structure and classification. The results that we present were obtained in [Bri09] and in [SS09], via two different approaches that are both valid over any base field. Here we mostly follow the former approach, with some simplifications.

5.1 Characterizations

Lemma 5.1.1 *Let G be an anti-affine group scheme. Then G is a connected commutative algebraic group. Moreover, G/H is anti-affine for any subgroup scheme H of G.*

PROOF: Clearly, G is connected. Also, since the scheme G/G_{red} is finite and $\mathcal{O}(G/G_{\text{red}}) \subset \mathcal{O}(G) = k$, we see that $G = G_{\text{red}}$. The commutativity of G follows from Theorem 1.2.1. For the final assertion, just note that $\mathcal{O}(G/H) = \mathcal{O}(G)^H \subset \mathcal{O}(G)$. $\qquad\square$

Proposition 5.1.2 *The following conditions are equivalent for an algebraic group G:*

(i) G is anti-affine.

(ii) Any rational finite-dimensional representation $\rho : G \to \mathrm{GL}_n$ is trivial.

(iii) Any non-trivial action of G on a variety has no fixed points.

PROOF: (i)$\Rightarrow$(ii) The matrix coefficients of ρ are global regular functions on G. As G is anti-affine these functions are just scalars, which implies that ρ is trivial.

(ii)$\Rightarrow$(i) By Theorem 1.2.1, there exists a largest anti-affine subgroup G_{ant} of G, and G/G_{ant} is affine. The quotient morphism $G \to G/G_{\mathrm{ant}}$ then gives a representation of G. By assumption, this representation has to be trivial, so by surjectivity of the quotient morphism we get that G/G_{ant} is trivial as required.

(i)$\Rightarrow$(iii) Let G act on a variety X and suppose this action α has a fixed point. Then the quotient of G by the kernel of α is affine by Proposition 2.1.6. But this quotient is also anti-affine, since G is so; hence α must be trivial.

(iii)$\Rightarrow$(ii) This is easy to see, since a rational representation $\rho : G \to \mathrm{GL}_n$ yields a G-action on the affine space $\mathbb{A}^n$ that fixes the origin. Hence this action has to be trivial. $\square$

Since the rigidity lemma for complete varieties extends to anti-affine schemes (Proposition 4.3.1), the classical properties of abelian varieties derived from that lemma also hold for anti-affine groups. Specifically, we have the following:

Lemma 5.1.3 *Let G be an anti-affine group, H a connected algebraic group and $f : G \to H$ a morphism (of varieties) such that $f(e_G) = e_H$. Then f is a group homomorphism and its image is contained in the center of H. Moreover, f is uniquely determined by the induced homomorphism of abelian varieties $\bar{f} : G/G_{\mathrm{aff}} \to H/H_{\mathrm{aff}}$.*

PROOF: The morphism

$$\varphi : G \times G \to H, \quad (x, y) \mapsto f(xy)f(x)^{-1}f(y)^{-1}$$

sends $G \times e_G$ to e_H. By Proposition 4.3.1, it follows that $\varphi(x,y) = \psi(y)$ for a unique morphism $\psi : G \to H$. In particular, $\psi(y) =$

$\varphi(e_G, y) = e_H$. Thus, $\varphi(x, y) = e_H$, *i.e.*, f is a group homomorphism. Its image is an anti-affine subgroup of H, and hence is central by Theorem 1.2.1. Finally, if $f' : G \to H$ is another morphism such that $f'(e_G) = e_H$ and $\bar{f} = \bar{f}'$, then the morphism $G \to H$, $g \mapsto f(g)f'(g)^{-1}$ has its image contained in H_{aff}. Since G is anti-affine, this morphism must be constant; hence $f = f'$ as required. $\square$

We shall see that these rigidity properties characterize anti-affine groups among connected algebraic groups. For this, we need some preliminaries on functors of morphisms; these are defined as follows (see [DG70, §I.2.7]).

Given two schemes X, Y, we assign to any scheme S the set

$$\underline{\mathrm{Hom}}(X, Y)(S) := \mathrm{Hom}_S(X \times S, Y \times S) \cong \mathrm{Hom}(X \times S, Y).$$

This yields a contravariant functor from schemes to sets, that we denote by $\underline{\mathrm{Hom}}(X, Y)$; note that $\underline{\mathrm{Aut}}(X)$ is a subfunctor of $\underline{\mathrm{Hom}}(X, X)$. If Y is a group scheme, then each $\mathrm{Hom}(X \times S, Y)$ has a group structure with operations being pointwise multiplication, resp. inverse; the neutral element is the constant morphism to e_Y. Moreover, $\mathrm{Hom}(X \times S, Y)$ contains $\mathrm{Hom}(S, Y) = Y(S)$ as a subgroup. One easily checks that $\underline{\mathrm{Hom}}(X, Y)$ is a group functor which contains Y as the subfunctor of "constant" morphisms; further, the tangent space to $\underline{\mathrm{Hom}}(X, Y)$ at the neutral element is isomorphic to $\mathcal{O}(X) \otimes \mathrm{Lie}(Y)$.

We may now state:

Proposition 5.1.4 *Let G and H be non-trivial connected algebraic groups. Then G is anti-affine if and only if the group functor $\underline{\mathrm{Hom}}(G, H)$ is representable by a scheme, locally of finite type. Under this assumption, we have $\underline{\mathrm{Hom}}(G, H) = H \times \mathrm{Hom}_{\mathrm{gp}}(G, H)$, where $\mathrm{Hom}_{\mathrm{gp}}(G, H)$ is viewed as a constant group scheme.*

PROOF: Assume that G is anti-affine. Let S be a connected scheme, and $f : G \times S \to H$ a morphism. Choose a point $s_0 \in S$. Then the morphism

$$G \times S \to H, \quad (g, s) \mapsto f(g, s)f(g, s_0)^{-1}$$

sends $G \times s_0$ to e_H. By Proposition 4.3.1, it follows that there exists a unique morphism $\varphi : S \to H$ such that $f(g, s) = \varphi(s)f(g, s_0)$. Further, the morphism

$$\psi : G \to H, \quad g \mapsto f(e_G, s_0)^{-1}f(g, s_0)$$

satisfies the assumptions of Lemma 5.1.3, and hence is a group homomorphism with central image. Since $f(g, s) = \varphi(s)f(e_G, s_0)\psi(g)$, we see that $\underline{\mathrm{Hom}}(G, H) = H\mathrm{Hom}_{\mathrm{gp}}(G, H)$. Moreover, this product is direct since the scheme-theoretic intersection $H \cap \mathrm{Hom}_{\mathrm{gp}}(G, H)$ is trivial.

Conversely, assume that the functor $\underline{\mathrm{Hom}}(G, H)$ is representable by a scheme locally of finite type. Then its tangent space at the neutral element must be a finite dimensional k-vector space, and hence so is $\mathcal{O}(G)$. Since G is connected, we get $\mathcal{O}(G) = k$ as required. $\qquad\square$

5.2   Structure

Let G be a connected commutative algebraic group. Recall from Chapter 1 that we get two exact sequences arising from Chevalley's theorem along with the structure of commutative linear algebraic groups:

$$0 \to U \to G_u := G/T \to A \to 0, \tag{5.1}$$

$$0 \to T \to G_s := G/U \to A \to 0, \tag{5.2}$$

where T is the torus and U is the connected commutative unipotent group such that $G_{\mathrm{aff}} = T \times U$.

Lemma 5.2.1 *With notation as above we have:*

(i) $G \cong G_s \times_A G_u$.

(ii) G *is anti-affine if and only if* G_s *and* G_u *are so.*

PROOF: (i) The homomorphisms $G \to G_s$ and $G \to G_u$ sit in a commutative square of group homomorphisms

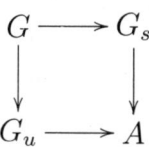

where the vertical arrows are principal T-bundles. Hence this square is cartesian.

(ii) If G is anti-affine, then so are its quotients G_s and G_u. To prove the converse, we show condition (ii) of Proposition 5.1.2, *i.e.*, given a rational representation $\rho : G \to \mathrm{GL}_n$, we show that ρ is trivial. Being a closed subgroup of GL_n, the image $\rho(G)$ is affine; since G is affine it follows that the quotient $\rho(G)/\rho(U)$ is also affine. By the definition of G_s, we get that the surjective map $\rho_s : G \to \rho(G)/\rho(U)$ induced by ρ, factors through G_s. However, G_s being anti-affine, its image under ρ_s has to be trivial, implying that $\rho(G) = \rho(U)$. Similarly, one can deduce from the anti-affineness of G_u that $\rho(G) = \rho(T)$. Thus, $\rho(G)$ being simultaneously unipotent and diagonalizable it has to be trivial, proving our claim. □

The point of the above lemma is that to study anti-affine groups it suffices to separately study those that are of the form G_s or G_u, *i.e.*, extensions of an abelian variety by a torus or by a connected commutative unipotent group.

5.3 Semi-abelian varieties

In this section, we assume that G is a connected algebraic group such that G_{aff} is a torus, denoted as T (then G is commutative by Proposition 3.1.1). In other words, we have an extension of commutative algebraic groups

$$0 \to T \to G \xrightarrow{\alpha} A \to 0. \qquad (5.3)$$

Our aims in this section are to classify such an extension by a homomorphism of abstract groups $c : \hat{T} \to \hat{A}(k)$, where $\hat{T}$ denotes the character group of T and $\hat{A}$ denotes the dual abelian variety, and to show that G is anti-affine if and only if c is injective. In order to define the map c we shall need a few observations.

Let $\mathcal{R} := \alpha_*(\mathcal{O}_G)$; this is a sheaf of algebras over A equipped with a T-action. This we may see as follows. Since $\alpha : G \to A$ is a principal T-bundle, it is locally trivial for the Zariski topology. Let $\{V_i\}$ be an open covering of A such that $\alpha^{-1}(V_i) \cong T \times V_i$ as a T-bundle over V_i. Then

$$\mathrm{H}^0(V_i, \mathcal{R}) = \mathrm{H}^0(V_i, \alpha_*(\mathcal{O}_G)) = \mathrm{H}^0(\alpha^{-1}(V_i), \mathcal{O}_G) \cong \mathcal{O}(T \times V_i)$$
$$\cong \mathcal{O}(T) \otimes \mathcal{O}(V_i)$$
$$(5.4)$$

and the action of T on $\mathcal{O}(T)$ via multiplication yields an action on $\mathrm{H}^0(V_i, \mathcal{R})$; these actions can be glued to give the desired action on $\mathcal{R}$.

Recall that

$$\mathcal{O}(T) = \bigoplus_{\lambda \in \hat{T}} k\lambda \qquad (5.5)$$

is the decomposition into T-weight spaces. This yields the decomposition of $\mathcal{R}$ into T-weight spaces,

$$\mathcal{R} = \bigoplus_{\lambda \in \hat{T}} \mathcal{R}_\lambda. \qquad (5.6)$$

We then note that

(i) Each $\mathcal{R}_\lambda$ is an invertible sheaf of $\mathcal{O}_A$-modules, hence defines a point of $\mathrm{Pic}(A)$.

(ii) The decomposition (5.6) is a grading of the sheaf of $\mathcal{O}_A$-algebras $\mathcal{R}$, i.e., $\mathcal{R}_\lambda \mathcal{R}_\mu \subset \mathcal{R}_{\lambda+\mu}$ for all $\lambda, \mu \in \hat{T}$,

(iii) Each $\mathcal{R}_\lambda$ is invariant under translations by A, i.e., it defines a point of $\mathrm{Pic}^\circ(A)$ by [Mil86, Proposition 10.1].

Indeed, (i) and (ii) follow from the isomorphisms (5.4) and the decomposition (5.5). For (iii), let $g \in G$. Then we have a canonical isomorphism (see [Har77, §III.9.3])

$$\tau^*_{\alpha(g)} \alpha_*(\mathcal{O}_G) \cong \alpha_* \tau^*_g(\mathcal{O}_G),$$

since τ_g is flat and we have a cartesian square

$$
\begin{array}{ccc}
G & \xrightarrow{\;\tau_g\;} & G \\
\downarrow{\scriptstyle\alpha} & & \downarrow{\scriptstyle\alpha} \\
A & \xrightarrow{\;\tau_{\alpha(g)}\;} & A
\end{array}
$$

Thus, $\tau_{\alpha(g)}^*(\mathcal{R}) \cong \mathcal{R}$ and this isomorphism commutes with the T-action. Taking λ-weight spaces we get the translation invariance of the $\mathcal{R}_\lambda$'s.

The above observations put together gives us a homomorphism $c : \hat{T} \to \hat{A}(k)$ which sends $\lambda \in \hat{T}$ to the class of $\mathcal{R}_\lambda$. We then have:

Proposition 5.3.1 *The map c as defined above classifies the extension (5.3).*

PROOF: In the case when $T = \mathbb{G}_m$ the statement follows from the isomorphism (see [Ser88, VII.16 Theorem 6])

$$
\mathrm{Ext}^1(A, \mathbb{G}_m) \xrightarrow{\;\cong\;} \hat{A}(k)
$$

that assigns to each extension (5.3) the class of the line bundle associated with the principal $\mathbb{G}_m$-bundle α. Indeed, via this isomorphism (called the Barsotti-Weil formula), the image of $1 \in \mathbb{Z} \cong \hat{T}$, under the map c, determines the extension.

In the general case, we have a canonical isomorphism $T \cong \mathrm{Hom}_{\mathrm{gp}}(\hat{T}, \mathbb{G}_m)$ which yields the desired isomorphism

$$
\mathrm{Ext}^1(A, T) \cong \mathrm{Hom}_{\mathrm{gp}}(\hat{T}, \hat{A}(k)).
$$

$\square$

Proposition 5.3.2 *With notations as above, G is anti-affine if and only if the map c is injective.*

PROOF: We have

$$
\mathcal{O}(G) = \mathrm{H}^0(A, \alpha_*(\mathcal{O}_G)) = \mathrm{H}^0(A, \mathcal{R}) = \bigoplus_{\lambda \in \hat{T}} \mathrm{H}^0(A, \mathcal{R}_\lambda). \tag{5.7}
$$

We claim that

$$\mathcal{R}_\lambda \cong \mathcal{O}_A \Leftrightarrow \mathrm{H}^0(A, \mathcal{R}_\lambda) \neq 0.$$

Indeed, the implication $\Rightarrow$ is obvious. On the other hand, if $\mathcal{R}_\lambda$ is not isomorphic to $\mathcal{O}_A$ but $\mathrm{H}^0(A, \mathcal{R}_\lambda) \neq 0$, then $\mathcal{R}_\lambda \cong \mathcal{O}_A(D)$ for a nonzero effective divisor D on A. We may then find an irreducible curve C in A such that C meets the support of D, but C is not contained in that support. Then $D \cdot C > 0$, which contradicts the fact that $\mathcal{R}_\lambda$ is algebraically trivial and hence has degree 0 on each curve. This completes the proof of the claim.

Now by (5.7), G is anti-affine if and only $\mathrm{H}^0(A, \mathcal{R}_\lambda) = 0$ for all $\lambda \neq 0$. In view of the claim, this is equivalent to $c(\lambda) \neq 0$ for all $\lambda \neq 0$. $\square$

5.4   Extensions of abelian varieties by vector groups

In this section, G denotes a connected commutative algebraic group with G_{aff} a (connected) unipotent group denoted as U; in other words, we have an extension of connected commutative algebraic groups

$$0 \to U \to G \xrightarrow{\alpha} A \to 0. \tag{5.8}$$

If $\mathrm{char}(k) > 0$ then the structure of U is rather complicated (see [Ser88, §VII.2]). However, in the light of the following proposition this does not pose a problem in the study of anti-affine groups.

Proposition 5.4.1 *Let* $\mathrm{char}(k)$ *be* $p > 0$. *Then* G *is anti-affine if and only if* $U = 0$.

PROOF: Since U is unipotent, we have $p^n U = 0$ for $n \gg 0$. So the push-out of the extension (5.8) by the multiplication map $p_U^n : U \to U$ has to be trivial, *i.e.*, we get the following commuting diagram of extensions

$$0 \longrightarrow U \longrightarrow G \longrightarrow A \longrightarrow 0$$
$$\Big\downarrow p^n \qquad \Big\downarrow \qquad \Big\downarrow \text{id}$$
$$0 \longrightarrow U \longrightarrow U \times A \longrightarrow A \longrightarrow 0.$$

Thus, we have shown that $p^n \text{Ext}^1(A, U) = 0$. By bilinearity of $\text{Ext}^1(A, U)$, it follows also that the pull-back extension via p_A^n is trivial, giving the following commuting diagram of extensions

$$0 \longrightarrow U \longrightarrow U \times A \longrightarrow A \longrightarrow 0$$
$$\Big\downarrow \text{id} \qquad \Big\downarrow \qquad \Big\downarrow p^n$$
$$0 \longrightarrow U \longrightarrow G \longrightarrow A \longrightarrow 0.$$

The morphism $U \times A \to G$ is finite and surjective, since p_A^n is so. Thus, $\mathcal{O}(U \times A)$ is a finite $\mathcal{O}(G)$-module. But $\mathcal{O}(U \times A) \cong \mathcal{O}(U) \otimes \mathcal{O}(A) \cong \mathcal{O}(U)$. Hence if G is anti-affine, then $\mathcal{O}(U)$ is a finite dimensional k-vector space, and hence $U = 0$. The other way implication is obvious. $\qquad \square$

If $\text{char}(k) = 0$, then U is a *vector group, i.e.,* a finite dimensional k-vector space regarded as an additive group. Moreover, in this case there exists a *universal extension by a vector group, i.e.,* an extension

$$0 \to H^1(A, \mathcal{O}_A)^* \to E(A) \to A \to 0 \qquad (5.9)$$

such that every extension (5.8) is the push-out of (5.9) by a unique *classifying map*

$$\gamma : H^1(A, \mathcal{O}_A)^* \to U.$$

In other words, we have a commuting diagram of extensions

$$0 \longrightarrow H^1(A, \mathcal{O}_A)^* \longrightarrow E(A) \longrightarrow A \longrightarrow 0 \qquad (5.10)$$
$$\Big\downarrow \gamma \qquad \Big\downarrow \delta \qquad \Big\downarrow \text{id}$$
$$0 \longrightarrow U \longrightarrow G \longrightarrow A \longrightarrow 0,$$

where γ is a linear map, and δ a group homomorphism. This results from the isomorphisms

$$\text{Ext}^1(A, U) \cong \text{H}^1(A, \mathcal{O}_A \otimes U) \cong \text{H}^1(A, \mathcal{O}_A) \otimes U$$
$$\cong \text{Hom}(\text{H}^1(A, \mathcal{O}_A)^*, U),$$

where the first isomorphism follows from [Ser88, VII.17 Theorem 7]. We refer to [MM74, Chap. 1] for details and further developments on universal extensions of abelian varieties by vector groups.

Proposition 5.4.2 *Let* $\text{char}(k) = 0$. *Then* G *is anti-affine if and only if the classifying map* γ *corresponding to the extension (5.8) is surjective.*

PROOF: The diagram (5.10) yields an isomorphism

$$\text{coker}(\gamma) \cong \text{coker}(\delta)$$

in view of the five-lemma. In particular, $\text{coker}(\delta)$ is affine. If G is anti-affine, then the morphism $G \to \text{coker}(\delta)$ must be constant, *i.e.*, δ is surjective; hence so is γ.

To prove the other way implication, it suffices to check that $E(A)$ (as described above) is anti-affine: indeed, $\delta : E(A) \to G$ is surjective since γ is so. By the Rosenlicht decomposition, we have

$$E(A) = \text{H}^1(A, \mathcal{O}_A)^* E(A)_{\text{ant}},$$

i.e., the map $\gamma_0 : \text{H}^1(A, \mathcal{O}_A)^* \to E(A)/E(A)_{\text{ant}}$ is surjective. We may view γ_0 as the classifying map for some extension G of A by the vector group $E(A)/E(A)_{\text{ant}}$. However, this extension has to be trivial since the quotient map $E(A) \to E(A)/E(A)_{\text{ant}}$ fits into a commuting diagram of extensions

$$
\begin{array}{ccccccccc}
0 & \longrightarrow & \text{H}^1(A, \mathcal{O}_A)^* & \longrightarrow & E(A) & \longrightarrow & A & \longrightarrow & 0 \\
& & \gamma_0 \downarrow & \swarrow & \downarrow & & \text{id} \downarrow & & \\
0 & \longrightarrow & E(A)/E(A)_{\text{ant}} & \longrightarrow & G & \longrightarrow & A & \longrightarrow & 0.
\end{array}
$$

This in turn implies that the extension G is trivial; equivalently, the classifying map, γ_0, is zero. But γ_0 is also surjective. Thus,

$E(A) = E(A)_{\text{ant}}$ as required.   □

To emphasize the analogy with semi-abelian varieties, we may also classify the extensions of A by a vector group U in terms of the transpose of the classifying map,

$$\gamma^t : U^* \to H^1(A, \mathcal{O}_A).$$

Then the extension is anti-affine if and only if the linear map γ^t is injective. Note that $H^1(A, \mathcal{O}_A)$ is canonically isomorphic to the Lie algebra of $\hat{A}$ (see [BLR90, Theorem 8.4.1]).

5.5   Classification

Let A be an abelian variety. A connected algebraic group G equipped with an isomorphism $G/G_{\text{aff}} \xrightarrow{\cong} A$ will be called a *group over A*.

From the discussions in the previous two sections, we have established the following.

Theorem 5.5.1   *(i) When* $\text{char}(k) > 0$, *the anti-affine groups G over A are classified by the subgroups Λ of $\hat{A}(k)$ which are free of finite rank.*

(ii) When $\text{char}(k) = 0$, *such groups are classified by the pairs* (Λ, V) *where Λ is as above and V is a linear subspace of* $\text{H}^1(A, \mathcal{O}_A)$.   □

Corollary 5.5.2 *Assume that k is the algebraic closure of a finite field. Then every anti-affine group is an abelian variety. Further, every connected algebraic group G is isomorphic to $(H \times A)/\Gamma$, where H is a connected affine algebraic group, A an abelian variety, and Γ a finite group scheme equipped with faithful homomorphisms $\Gamma \hookrightarrow Z(H)$ and $\Gamma \hookrightarrow A$; then H, A and Γ are uniquely determined by G.*

PROOF: Since k is the union of its finite subfields, the group $\hat{A}(k)$ is torsion. Together with the above theorem, this proves the first assertion. The second assertion follows from the Rosenlicht decomposition (Theorem 1.2.4): here $H = G_{\text{aff}}$, $A = G_{\text{ant}}$ and $\Gamma = G_{\text{aff}} \cap G_{\text{ant}}$.
□

In the next two chapters, we shall need to slightly generalize the setting of anti-affine groups over an abelian variety A; namely, we shall encounter exact sequences of algebraic groups

$$0 \to H \to G \to A \to 0, \tag{5.11}$$

where G is an anti-affine group, and H an affine subgroup scheme. Such an exact sequence will be called an *anti-affine extension of A*.

We now see how to reduce the classification of anti-affine extensions to that of anti-affine groups over abelian varieties B equipped with an isogeny $B \to A$. Given an exact sequence (5.11), one easily checks that $G_{\text{aff}} = H^o_{\text{red}}$; in particular, the quotient $N := H/G_{\text{aff}}$ is finite. Thus, the abelian variety $A(G) := G/G_{\text{aff}}$ (the Albanese variety of G) is equipped with an isogeny $A(G) \twoheadrightarrow A$ with kernel N, and G is of course an anti-affine group over $A(G)$. By duality for abelian varieties (see [Mil86, Theorem 11.1]), the isogenies $B \to A$ with (scheme-theoretic) kernel N, where B is an abelian variety, are classified by the homomorphisms $\hat{N} \to \hat{A}$ with a trivial kernel, where $\hat{N}$ denotes the Cartier dual of the finite commutative group scheme N. It follows that *the anti-affine extensions of A are classified by the pairs consisting of a finite subgroup scheme $\Gamma \subset \hat{A}$, and an anti-affine group over the abelian variety dual to $\hat{A}/\Gamma$.*

To classify anti-affine extensions of A, one may also adapt the arguments of Sections 5.3 and 5.4. We now state the result obtained in this way, referring to [Bri12, Section 3.3] for details:

Theorem 5.5.3   *(i) When* $\text{char}(k) > 0$, *the anti-affine extensions of A are classified by the pairs (Λ, Γ), where Λ is a finitely generated subgroup of $\hat{A}(k)$, and $\Gamma \subset \hat{A}$ is a subgroup scheme supported at the origin.*

(ii) When char(k) = 0, such extensions are classified by the triples (Λ, Γ, V), where Λ, Γ are as above and V is a linear subspace of $\mathrm{H}^1(A, \mathcal{O}_A)$. □

The commutative affine group scheme H can be recovered from the data in the above theorem: write $H = D \times U$ where D denotes the diagonalizable part of H, and U the unipotent part. In case (i), U must be finite, and Γ is its Cartier dual; in case (ii), U is a vector group, and V is the dual of the vector space U. In both cases, Λ is the character group of the diagonalizable group scheme D.

Finally, we present a geometric application of the classification of anti-affine groups, to be used in the next chapter.

Proposition 5.5.4 Let X be a variety. Then X has a largest anti-affine group of automorphisms, that we denote as $\mathrm{Aut}_{\mathrm{ant}}(X)$. Moreover, $\mathrm{Aut}_{\mathrm{ant}}(X)$ centralizes every connected group scheme of automorphisms of X.

PROOF: Let G be an anti-affine subgroup of $\mathrm{Aut}(X)$, and H be a connected subgroup scheme of $\mathrm{Aut}(X)$. Consider the morphism

$$\varphi : G \times H \times X, \quad (g, h, x) \mapsto ghg^{-1}h^{-1}x.$$

Then $\varphi(G \times e_H \times x) = x$ for any $x \in X$. By Proposition 4.3.1, it follows that $\varphi(g, h, x) = \psi(h, x)$ for some morphism $\psi : H \times X \to X$. Thus, $\psi(h, x) = \varphi(e_g, h, x) = x$, and hence G centralizes H.

Now, suppose H is an anti-affine group of automorphisms of X. Then we just saw that the direct product $G \times H$ acts on X; moreover, the product $GH \subset \mathrm{Aut}(X)$ (the quotient of $G \times H$ by the scheme-theoretic kernel of the action) is an anti-affine group of automorphisms; in particular, GH is connected. Thus, to show the existence of $\mathrm{Aut}(X)_{\mathrm{ant}}$, it suffices to bound the dimension of any such group G. Let $G_{\mathrm{aff}} = T \times U$ and $G/G_{\mathrm{aff}} = A$ as above. Then T acts on the variety X with a trivial scheme-theoretic kernel, and hence by Lemma 5.5.5 the isotropy group T_x of a general point $x \in X$ is trivial as well. Thus, $\dim(T) = \dim(T \cdot x) \leq \dim(X)$. Also, G_x is affine by Corollary 2.1.9, and hence $(G_x)^o_{\mathrm{red}}$ is contained in G_{aff}. Thus,

$$\dim(A) = \dim(G) - \dim(G_{\text{aff}}) \leq \dim(G) - \dim(G_x)$$
$$= \dim(G \cdot x) \leq \dim(X).$$

Finally, $\dim(U) \leq \dim(A)$ by Proposition 5.4.2. Thus, $\dim(G) = \dim(T) + \dim(U) + \dim(A) \leq 3 \dim(X)$. $\qquad\square$

Lemma 5.5.5 *Let T be a torus of automorphisms of a variety X. Then there exists a dense open subset U of X such that the isotropy subgroup scheme T_x is trivial for any $x \in U$.*

PROOF: Replacing X with a T-stable open subset, we may assume that X is normal. Then, by a classical theorem of Sumihiro [Sum74], X is covered by affine T-stable open subsets; thus, we may assume that X itself is affine. Then, since the action of T is scheme-theoretic faithful, so is the action on $\mathcal{O}(X)$; equivalently, the weights of T on $\mathcal{O}(X)$ generate the character group $\hat{T}$. Thus, we may choose finitely many such weights, say $\lambda_1, \ldots, \lambda_n$, that generate $\hat{T}$. Next, choose $f_i \in \mathcal{O}(X)_{\lambda_i}$ for $i = 1, \ldots, n$, and let U denote the open subset of X where $f_i \neq 0$ for all i. Then U is clearly T-stable and non-empty. Moreover, for any $x \in U$, each f_i restricts to a non-zero $g_i \in \mathcal{O}(\overline{T.x})_{\lambda_i}$, and hence the scheme-theoretic kernel of the T-action on $\overline{T.x}$ is trivial. It follows that the group scheme T_x is trivial as well. $\qquad\square$

Remark 5.5.6 (i) If X is affine, then $\text{Aut}_{\text{ant}}(X)$ is trivial. (Indeed, this group acts faithfully on $\mathcal{O}(X)$, so that the assertion follows from Proposition 5.1.2).

(ii) If G is a connected algebraic group (viewed as a variety), then $\text{Aut}_{\text{ant}}(G) = G_{\text{ant}}$. (Indeed, $\text{Aut}_{\text{ant}}(G)$ centralizes G acting by right multiplication, and hence $\text{Aut}_{\text{ant}}(G)$ is identified with a subgroup of G acting by left multiplication).

Chapter 6

Homogeneous vector bundles over abelian varieties

In this chapter, we present a proof of the results of Matsushima, Morimoto, Miyanishi and Mukai stated in Theorem 1.5.1. For this, we follow the approach of [Bri12], based on techniques of algebraic groups.

We begin by recalling some general notions and results on principal G-bundles over arbitrary schemes, where G stands for an affine algebraic group. Then we show that the equivariant automorphisms of a principal bundle over any proper scheme form a group scheme, locally of finite type (Theorem 6.2.1); moreover, the bundle automorphisms form an affine subgroup scheme of finite type (Proposition 6.2.3).

When G is the general linear group GL_n, the principal G-bundles are in one-to-one correspondence with the vector bundles of rank n; we describe the induced isomorphisms of equivariant automorphism groups. For line bundles, we obtain another view of Mumford's Theta group.

Next, we obtain a classification of homogeneous vector bundles over an abelian variety A, in terms of anti-affine extensions and associated vector bundles (Theorem 6.4.1). From this, we derive the announced proof of Theorem 1.5.1, and also an alternative approach to the Mukai correspondence between homogeneous vector bundles on A and coherent sheaves with finite support on $\hat{A}$.

6.1   Principal bundles

Throughout this chapter, we denote by G an affine algebraic group.

Definition 6.1.1 *A principal bundle (also known as a torsor) under G is a morphism of schemes $\pi : X \to Y$ which satisfies the following conditions:*

(i) *X is equipped with an action α of G such that π is G-invariant.*

(ii) *π is faithfully flat.*

(iii) *The diagram*

$$
\begin{array}{ccc}
G \times X & \xrightarrow{p_2} & X \\
{\scriptstyle \alpha}\downarrow & & \downarrow{\scriptstyle \pi} \\
X & \xrightarrow{\pi} & Y
\end{array}
$$

is cartesian, where α denotes the action, and p_2 the projection.

Under these assumptions, we say for simplicity that π is a G-*bundle*. Then π is smooth and its fibres are isomorphic to G; also, π is a geometric quotient by G, and hence a categorical quotient (see [MFK94, Proposition 0.1]).

In the above definition, conditions (ii) and (iii) may be replaced with:

(iv) *For any point $y \in Y$, there exist an open subset V of Y containing y and a faithfully flat morphism $f : U \to V$ such that the pull-back morphism $X \times_Y V \to V$ is isomorphic to the trivial bundle $p_U : G \times U \to U$ as a G-scheme over U.*

The morphism π is also said to be a *locally trivial bundle for the fppf topology* (under our standing assumption of finiteness for schemes). Since G is assumed to be affine, any such bundle is also locally trivial for the étale topology, *i.e.*, we may replace 'faithfully flat' with 'étale' in condition (iv). In fact, π is *locally isotrivial, i.e.,* we may take f to be a finite étale covering (see [Gro60, pp. 27–28], [Ray70, Lemme XIV 1.4]). This does not extend to an arbitrary group G, as shown by the following:

Example 6.1.2 Let E be an elliptic curve, and $\pi : X \to C$ the E-bundle over a nodal curve constructed in Example 3.1.5. Let V be an open neighborhood of the singular point of C. We show that there exists no finite étale covering $f : U \to V$ such that the pull-back bundle $X \times_V U$ is trivial.

We may assume that f is a Galois covering. Then its Galois group Γ acts on $X \times_V U$ by automorphisms which commute with the action of E. If $X \times_V U \cong E \times U$ as E-bundles over U, then we get an action of Γ on $E \times U$ which lifts the given action on U, and commutes with the action of E by translations on itself. Such an action must be of the form $\gamma \cdot (p, u) = (p + \rho(\gamma), \gamma \cdot u)$ with obvious notations, where $\rho : \Gamma \to E(k)$ is a group homomorphism. Thus, $\rho(\Gamma)$ is contained in the n-torsion subgroup scheme E_n for some n. But then the projection $E \times U \to E$ yields an E-equivariant morphism $(E \times U)/\Gamma \to E/\rho(\Gamma) \to E/E_n$. Since $(E \times U)/\Gamma \cong (X \times_V U)/\Gamma \cong \pi^{-1}(V)$, we get an equivariant morphism $\pi^{-1}(V) \to E/E_n$. But such a morphism does not exist, as seen in Example 3.1.5.

Given a G-bundle $\pi : X \to Y$ and a scheme Z equipped with a G-action, the *associated bundle* is a scheme W equipped with morphisms $q : X \times Z \to W$, $\pi_Z : W \to Y$ such that the square

$$\begin{array}{ccc} X \times Z & \xrightarrow{p_1} & X \\ {\scriptstyle q}\downarrow & & \downarrow{\scriptstyle \pi} \\ W & \xrightarrow{\pi_Z} & Y \end{array}$$

is cartesian, where p_1 denotes the projection. Then q is a G-bundle relative to the diagonal action of G on $X \times Z$, and hence W is uniquely defined; we denote it as $X \times^G Z$. Moreover, π_Z is locally trivial for the fppf (or étale) topology, with fibre Z.

The associated bundle need not exist in general. We now present a simple criterion for its existence, and describe its sections:

Lemma 6.1.3 *Let* $\pi : X \to Y$ *be a G-bundle, and Z a scheme equipped with an action of G.*

(i) *The associated bundle* $\pi_Z : X \times^G Z \to Y$ *exists if Z admits a G-equivariant embedding into the projectivization of a finite-dimensional G-module.*

(ii) *Given a subgroup* $H \subset G$, *the morphism* π *factors as* $X \xrightarrow{\phi} Z \xrightarrow{\psi} Y$ *where* ϕ *is an H-bundle (obtained from* π *by reduction of structure group), and* ψ *a smooth morphism with fibres isomorphic to* G/H. *If H is a normal subgroup, then* ψ *is a G/H-bundle.*

(iii) *The set of sections of* π_Z *may be identified with* $\mathrm{Hom}^G(X, Z)$ *(the G-equivariant morphisms from X to Z).*

PROOF: (i) follows from a descent argument; specifically, from [MFK94, Proposition 7.1] applied to the morphism p_1.

(ii) By a theorem of Chevalley (see [Spr09, Theorem 5.5.3]), the homogeneous G-variety $Z := G/H$ admits a G-equivariant embedding into the projectivization of a G-module; thus, the associated bundle π_Z exists. We now take for ϕ the morphism $X \to X \times^G Z$ induced by the morphism $X \to X \times Z$, $x \mapsto (x, H)$, and take $\psi = \pi_Z$. Then ϕ is H-invariant, ψ is smooth with fibres isomorphic to Z, and $\psi \circ \phi = \pi$. Moreover, after the faithfully flat base change $\pi : X \to Y$, the morphism ϕ is identified with $G \times X \to G/H \times X$, $(g, x) \mapsto (gH, x)$ which clearly is an H-bundle. Thus, ϕ is an H-bundle as well. The final assertion is proved along similar lines.

(iii) Let $s : Y \to X \times^G Z$ be a section of π_Z. We may view s as a G-invariant morphism $X \to X \times^G Z$ such that the morphism $\mathrm{id} \times s : X \to X \times (X \times^G Z)$ factors through a morphism $X \to X \times_Y (X \times^G Z)$. In view of the above cartesian square, this yields a section $\sigma : X \to X \times Z$ of p_1, i.e., a morphism $f : X \to Z$, which must be equivariant since s is G-invariant. Conversely, any $f \in \mathrm{Hom}^G(X, Z)$ yields a G-invariant morphism $q \circ (\mathrm{id} \times f) : X \to X \times^G Z$ and hence a morphism $s : Y \to X \times^G Z$; one checks that the assignments $s \mapsto f$ and $f \mapsto s$ are inverses to each other.   □

In particular, $X \times^G Z$ exists whenever Z is affine. For instance, if Z is a rational finite-dimensional G-module, denoted as V, then the associated bundle

$$\pi_V : E_V := X \times^G V \to Y$$

exists, and is actually a vector bundle on Y (since it becomes the trivial vector bundle after pull-back by the faithfully flat morphism $\pi : X \to Y$).

The *adjoint bundle* is the associated bundle $\pi_G : X \times^G G \to Y$, where G acts on itself by conjugation. By the above lemma and the assumption that G is affine, this bundle exists; its set of sections is identified with $\text{Hom}^G(X, G)$. The latter has a group structure given by pointwise multiplication. We now identify this group with the group $\text{Aut}_Y^G(X)$ of *bundle automorphisms*, *i.e.*, of G-equivariant automorphisms ϕ of X such that the following diagram commutes:

$$
\begin{array}{ccc}
X & \xrightarrow{\phi} & X \\
\pi \downarrow & & \downarrow \pi \\
Y & \xrightarrow{\text{id}} & Y.
\end{array}
$$

For this, note that each $f \in \text{Hom}^G(X, G)$ defines a morphism

$$\phi : X \to X, \text{ given by } x \mapsto f(x) \cdot x,$$

which is readily seen to be a bundle automorphism. Moreover, the assignment $f \mapsto \phi$ defines a group homomorphism

$$\theta : \text{Hom}^G(X, G) \to \text{Aut}_Y^G(X). \tag{6.1}$$

Lemma 6.1.4 *With the notations as above, θ is an isomorphism.*

PROOF: Clearly, θ is injective. To show the surjectivity, just note that every bundle automorphism ϕ may be viewed as a G-equivariant section of the projection $p_2 : X \times_Y X \to X$, where G acts diagonally on $X \times_Y X$. But p_2 is identified with the projection $p_1 : G \times X \to X$ in view of condition (ii) of Definition 6.1.1, and this identifies the above G-action with the action on $G \times X$ via conjugation on G and the given action on X. Thus, an equivariant section of p_2 is an equivariant morphism $X \to G$. $\qquad\square$

Remark 6.1.5 *The center Z of G acts on X by bundle automorphisms; in fact, this identifies Z to a central subgroup scheme of $\mathrm{Aut}_Y^G(X)$. The pre-image under θ of this subgroup scheme consists of the constant morphisms from X to Z.*

If G is commutative, i.e., $G = Z$, then the adjoint bundle is trivial, and $\mathrm{Hom}^G(X, G)$ is identified with $\mathrm{Hom}(Y, G)$, the group of Y-points of G. In concrete words, every bundle automorphism is of the form $x \mapsto f(\pi(x)) \cdot x$ for a unique morphism $f : Y \to G$.

6.2   Equivariant automorphisms

Let $\pi : X \to Y$ be a G-bundle. Denote by $\mathrm{Aut}^G(X)$ the group of G-equivariant automorphisms of the G-variety X. (Note that these are not necessarily bundle automorphisms). If $f \in \mathrm{Aut}^G(X)$ then $\pi \circ f : X \to Y$ is a G-invariant morphism, hence factors via the categorical quotient $X /\!/ G = Y$; the resulting morphism $\phi : Y \to Y$ is readily seen to be an isomorphism. So there exists a natural homomorphism $\pi_* : \mathrm{Aut}^G(X) \to \mathrm{Aut}(Y)$ which sits in the following exact sequence of groups:

$$1 \to \mathrm{Aut}_Y^G(X) \to \mathrm{Aut}^G(X) \xrightarrow{\pi_*} \mathrm{Aut}(Y). \qquad (6.2)$$

The image of π_* consists of those automorphisms ψ of Y such that $\psi^*(X) \cong X$ as G-bundles over Y.

As was mentioned in Chapter 4, the group functor $\underline{\mathrm{Aut}}(Y)$ given by $S \mapsto \mathrm{Aut}_S(Y \times S)$ is represented by a group scheme which is locally of finite type, whenever Y is proper. In particular, the group functor $\underline{\mathrm{Aut}}^\circ(Y)$ is an honest group scheme of finite type. We now consider the group functor $\underline{\mathrm{Aut}}^G(X)$ given by $S \mapsto \mathrm{Aut}_S^G(X \times S)$ where G acts on X by the given action and acts trivially on S.

Theorem 6.2.1 *With notations as above, assume that Y is proper. Then $\underline{\mathrm{Aut}}^G(X)$ is represented by a group scheme, locally of finite type.*

PROOF: The idea of the proof is to compactify X in an equivariant way such that all the G-automorphisms of X will extend to the compactification.

In order to compactify X, we first compactify G. Let $\bar{G}$ denote a $G \times G$-equivariant compactification of G constructed as in the proof of Proposition 3.1.1(iv): we embed G as a subgroup of GL_n which in turn embeds into the projective space $\mathbb{P}(\mathbb{M}_n \oplus k)$, and then take the closure of G in $\mathbb{P}(\mathbb{M}_n \oplus k)$. Since $\mathbb{M}_n \oplus k$ is a $G \times G$-module, we may apply Lemma 6.1.3 and form the associated bundle

$$\bar{\pi} : X \times^G \bar{G} \to Y$$

with fibre $\bar{G}$. Note that $X \times^G \bar{G}$ contains $X \times^G G$ as an open subscheme, isomorphic to X via the morphism $(x, g) \mapsto g \cdot x$; this isomorphism is equivariant relative to the action of G on $X \times^G G$ via right multiplication on G, and the given action on X. We thus get the following commuting diagram

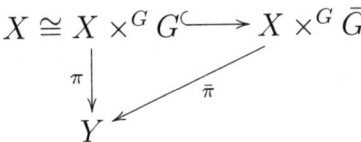

The morphism $\bar{\pi}$ is projective, since $\bar{G} \hookrightarrow \mathbb{P}(\mathbb{M}_n \oplus k)$ and the map $X \times^G \mathbb{P}(\mathbb{M}_n \oplus k) \to Y$ is the projectivization of a vector bundle over Y. Now, since Y is proper the scheme $\bar{X} = X \times^G \bar{G}$ also is proper. Further, X is open and dense in $\bar{X}$, since G is open and dense in $\bar{G}$. Also, the right action of G on $\bar{G}$ yields an action on $\bar{X}$ which extends the given G-action. With respect to this G-action on $\bar{X}$, any G-automorphism of X can be extended uniquely to a G-automorphism of $\bar{X}$. This is done as follows.

Let $\phi \in \mathrm{Aut}^G(X)$. Consider the morphism

$$\phi \times \mathrm{id} : X \times \bar{G} \to X \times \bar{G}.$$

This is a $G \times G$-equivariant automorphism, for the action on $X \times \bar{G}$ given by $(g, h)(x, y) = (gx, (g, h) \cdot y)$. Since the left action of G was used to construct the associated bundle $\bar{X}$, the automorphism $\phi \times \mathrm{id}$ being equivariant for this action descends to an automorphism $\bar{\phi}$ on $\bar{X}$. Moreover, $\bar{\phi}$ is G-equivariant with respect to the above action of G on the right. Also, note that $\phi \times \mathrm{id}$ preserves $X \times G$; thus $\bar{\phi}$ preserves X. Once again, using the isomorphism $X \times^G G \cong X$, we see that $\bar{\phi}\,|_X = \phi$.

Any automorphism of $\bar{X}$ constructed as above stabilizes X and hence also the boundary $\partial\bar{X} = \bar{X}\backslash X$. Thus, we get a homomorphism

$$\rho : \mathrm{Aut}^G(X) \to \mathrm{Aut}^G(\bar{X}, \partial\bar{X}).$$

Moreover, any automorphism of $\bar{X}$ that stabilizes $\partial\bar{X}$, stabilizes X. Hence the restriction to X gives the inverse of ρ. Thus, ρ is an isomorphism of groups.

Using the same argument it can proved that ρ extends to an isomorphism of group functors. Here the functor $\underline{\mathrm{Aut}}^G(\bar{X}, \partial\bar{X})$ is given by

$$S \mapsto \mathrm{Aut}^G_S(\bar{X} \times S, \partial\bar{X} \times S),$$

where G acts trivially on S; the functor $\underline{\mathrm{Aut}}^G(X)$ is defined similarly; since the conditions of G-equivariance and of preservation of $\partial\bar{X}$ are closed conditions, $\underline{\mathrm{Aut}}^G(\bar{X}, \partial\bar{X})$ is a closed subfunctor of $\underline{\mathrm{Aut}}(\bar{X})$. That is, we have

$$\underline{\mathrm{Aut}}^G(X) \xrightarrow{\ \rho\ } \underline{\mathrm{Aut}}^G(\bar{X}, \partial\bar{X}) \lhook\joinrel\longrightarrow \underline{\mathrm{Aut}}(\bar{X})\,,$$

an exact sequence of group functors. The rightmost term is representable by a group scheme, locally of finite type since $\bar{X}$ is complete. The functor $\underline{\mathrm{Aut}}^G(\bar{X}, \partial\bar{X})$, being a closed subfunctor of a representable functor, is itself representable; so is $\underline{\mathrm{Aut}}^G(X)$. $\square$

Remark 6.2.2 We can also define the group functor of bundle automorphisms $\underline{\mathrm{Aut}}^G_Y(X)$, by assigning to each scheme S the group $\mathrm{Aut}^G_{Y \times S}(X \times S)$. Then the exact sequence (6.2) extends to an exact sequence of group functors

$$1 \to \underline{\mathrm{Aut}}^G_Y(X) \to \underline{\mathrm{Aut}}^G(X) \to \underline{\mathrm{Aut}}(Y).$$

In view of the above theorem, this yields an exact sequence of group schemes.

In fact, the functor $\underline{\mathrm{Aut}}_Y^G(X)$ is represented by not just any group scheme but an affine group scheme of finite type. A fact that we shall see next.

With notations as above, consider the group functor $\underline{\mathrm{Hom}}^G(X, G)$ given by $S \mapsto \mathrm{Hom}_S^G(X \times S, G \times S) = \mathrm{Hom}^G(X \times S, G)$. We can define a morphism of group functors $\theta : \underline{\mathrm{Hom}}^G(X, G) \to \underline{\mathrm{Aut}}_Y^G(X)$ analogous to the homomorphism (6.1).

Proposition 6.2.3 *With notations as above, θ is an isomorphism. If Y is proper, then $\underline{\mathrm{Aut}}_Y^G(X)$ is represented by an affine group scheme of finite type.*

PROOF: The first assertion is proved by the argument of Lemma 6.1.4. To show the second assertion, we may view G as a closed subgroup of some SL_m (as in the proof of Lemma 2.1.1) and hence as a closed subvariety of the affine space $\mathbb{M}_m$, such that the conjugation action of G on itself extends to an action of $\mathbb{M}_m$. Then $\underline{\mathrm{Hom}}^G(X, G)$ is a closed subfunctor of $\underline{\mathrm{Hom}}^G(X, \mathbb{M}_m)$. The latter can be identified with the functor of points of the vector space of global sections of the associated vector bundle $E := X \times^G \mathbb{M}_m \to Y$, where this vector space, $\mathrm{H}^0(Y, E)$, is viewed as an affine scheme. But $\mathrm{H}^0(Y, E)$ is finite dimensional, since Y is proper. Thus, $\underline{\mathrm{Hom}}^G(X, G)$ is represented by an affine scheme of finite type. $\square$

6.3 Automorphisms of vector bundles

We now replace principal bundles with vector bundles.

Let Y be a scheme and $p : E \to Y$ a vector bundle. The multiplicative group $\mathbb{G}_m$ acts by scalars on the fibres of p, thus giving an action on E. Then, Y is the categorical quotient for this action, since this is true locally. Thus, any $\mathbb{G}_m$-equivariant automorphism of the scheme E induces an automorphism of Y and we get the following exact sequence

$$1 \to \mathrm{Aut}_Y^{\mathbb{G}_m}(E) \to \mathrm{Aut}^{\mathbb{G}_m}(E) \xrightarrow{p_*} \mathrm{Aut}(Y), \qquad (6.3)$$

where $\mathrm{Aut}_Y^{\mathbb{G}_m}(E)$ denotes the group of $\mathbb{G}_m$-equivariant automorphisms of E which induces the identity on Y.

Lemma 6.3.1 *With the above notation, $\mathrm{Aut}_Y^{\mathbb{G}_m}(E)$ consists of the vector bundle automorphisms of E. Moreover, the image of p_* consists of those $\psi \in \mathrm{Aut}(Y)$ such that $\psi^*(E) \cong E$ as vector bundles over Y.*

PROOF: Let $\varphi \in \mathrm{Aut}_Y^{\mathbb{G}_m}(E)$. Let $V \subset Y$ be an open subscheme such that $p^{-1}(V) \cong V \times k^n$ as vector bundles over V. Then $\varphi|_{p^{-1}(V)}$ is given by $(y, z) \mapsto (y, \Phi(y, z))$, where $\Phi : V \times k^n \to k^n$ satisfies

$$\Phi(y, tz) = t\Phi(y, z) \qquad (\star)$$

for $y \in V$, $z \in k^n$, $t \in k^*$. Since $\mathcal{O}(V \times k^n) = \mathcal{O}(V) \otimes_k \mathcal{O}(k^n) = \mathcal{O}(V)[x_1, \ldots, x_n]$, we have

$$\Phi(y, z) = \sum_{i=1}^{n} \sum_{a_1, \ldots, a_n} \Phi_{a_1, \ldots, a_n, i}(y) z_1^{a_1} \cdots z_n^{a_n} e_i,$$

where $(e_1, \ldots, e_n)$ denotes the standard basis of k^n. The homogeneity condition $(\star)$ forces $\Phi_{a_1, \ldots, a_n, i} = 0$ unless $a_1 + \cdots + a_n = 1$, *i.e.*, Φ is linear in z. Thus, φ is a vector bundle automorphism. Conversely, every vector bundle automorphism is obviously in $\mathrm{Aut}_Y^{\mathbb{G}_m}(E)$.

Next, let $\psi = p_*(\varphi)$, where $\varphi \in \mathrm{Aut}^{\mathbb{G}_m}(E)$. Then φ factors as an isomorphism $\gamma : E \to \psi^*(E)$ of schemes over Y, followed by the isomorphism $\psi^*(E) \to E$ induced by ψ. Moreover, γ is $\mathbb{G}_m$-equivariant. Arguing as above, we get that γ is an isomorphism of vector bundles. Conversely, every such isomorphism $\gamma : E \to \psi^*(E)$ yields a $\mathbb{G}_m$-equivariant automorphism of E lifting ψ. $\qquad\square$

A vector bundle endomorphism of E may be viewed as a global section of $\underline{End}(E)$, the sheaf of endomorphisms of E. Under this identification, the set of vector bundle automorphisms of E corresponds to a subset, $\mathrm{H}^0(Y, \underline{GL}(E))$, of the vector space $\mathrm{H}^0(Y, \underline{End}(E))$; the latter is an associative algebra, and $\mathrm{H}^0(Y, \underline{GL}(E))$ is its group of invertible elements. If Y is proper, then $\mathrm{H}^0(Y, \underline{End}(E))$ is finite dimensional and we note that

$H^0(Y, \underline{GL}(E))$ is a principal open subset of this affine space, hence is itself affine.

We know that vector bundles of rank n over Y are in one-to-one correspondence with GL_n-bundles over Y. For $G = GL_n$ and a G-bundle $\pi : X \to Y$, we have the exact sequence (6.2). The corresponding vector bundle $p : E \to Y$ yields the exact sequence (6.3). We shall now see that these two sequences can be identified with each other.

Under the above correspondence, E is the associated vector bundle $X \times^G k^n$ so that we have a natural homomorphism

$$\gamma : \text{Aut}^G(X) \to \text{Aut}^{\mathbb{G}m}(E).$$

Indeed, if $\phi \in \text{Aut}^G(X)$, the G-automorphism $\phi \times \text{id}$ on $X \times k^n$ descends to an automorphism $\gamma(\phi)$ of E. Under γ, the group of bundle automorphisms $\text{Aut}^G_Y(X)$ maps to the group of vector bundle automorphisms $\text{Aut}^{\mathbb{G}m}_Y(E)$, in view of Lemma 6.3.1. In fact, γ is an isomorphism of groups since we have

$$\text{Aut}^G_Y(X) \cong \text{Hom}^G(Y, G) \hookrightarrow \text{Hom}^G(Y, \mathbb{M}_n) \cong H^0(Y, \underline{End}(E))$$

where the first isomorphism is given by Lemma 6.1.4, the inclusion arises from the inclusion $G \hookrightarrow \mathbb{M}_n$ and the last isomorphism from the identification of $\underline{End}(E)$ with the associated vector bundle $X \times^G \mathbb{M}_m$; further, $\text{Aut}^G_Y(X)$ is sent isomorphically to $H^0(Y, \underline{GL}(E))$ under the above morphism.

We thus have the following commuting diagram:

$$
\begin{array}{ccccccc}
1 & \longrightarrow & \text{Aut}^G_Y(X) & \longrightarrow & \text{Aut}^G(X) & \xrightarrow{\pi_*} & \text{Aut}(Y) \\
& & \downarrow{\scriptstyle \imath} & & \downarrow{\scriptstyle \gamma} & & \| \\
1 & \longrightarrow & \text{Aut}^{\mathbb{G}m}_Y(E) & \longrightarrow & \text{Aut}^{\mathbb{G}m}(E) & \xrightarrow{p_*} & \text{Aut}(Y).
\end{array}
\qquad (6.4)
$$

Also, the maps π_* and p_* have the same image, since for any $\psi \in Aut(Y)$, we have that $\psi^*(E) \cong E$ as vector bundles if and only if $\psi^*(X) \cong X$ as G-bundles. Applying five-lemma to the above diagram we hence conclude that γ is an isomorphism. Thus we can identify both the automorphism groups $\text{Aut}^G(X)$ and $\text{Aut}^{\mathbb{G}m}(E)$;

the same holds for the corresponding group schemes by similar arguments. In view of Theorem 6.2.1 and of the above description of bundle automorphisms, this yields:

Proposition 6.3.2 *Let* $p : E \to Y$ *be a vector bundle over a proper scheme. Then the group functor of equivariant automorphisms,* $\underline{\mathrm{Aut}}^{\mathbb{G}_m}(E)$, *is represented by a group scheme, locally of finite type. Moreover, the group functor of bundle automorphisms,* $\underline{\mathrm{Aut}}_Y^{\mathbb{G}_m}(E)$, *is represented by a connected affine algebraic group.*   □

Let us specialize to the case of line bundles (*i.e.*, vector bundles of rank 1). Then $G = \mathbb{G}_m$, and the group of bundle automorphisms is

$$\mathrm{Hom}^{\mathbb{G}_m}(X, \mathbb{G}_m) = \mathrm{Hom}(Y, \mathbb{G}_m) = \mathbb{G}_m(Y) = \mathcal{O}(Y)^*,$$

the group of invertible elements of $\mathcal{O}(Y)$. If Y is a complete variety, then $\mathcal{O}(Y)^* = k^*$; thus, for a line bundle L, the exact sequence (6.3) gives

$$1 \to \mathbb{G}_m \to \mathrm{Aut}^{\mathbb{G}_m}(L) \to \mathrm{Aut}(Y).$$

Let H be a group scheme acting on the scheme Y; so we have a homomorphism $\rho : H \to \mathrm{Aut}(Y)$. Let $\mathcal{G}(H)$ denote the fibre product of H and $\mathrm{Aut}^{\mathbb{G}_m}(L)$ over $\mathrm{Aut}(Y)$, *i.e.*, we have a cartesian diagram:

$$
\begin{array}{ccccc}
1 \longrightarrow & \mathbb{G}_m & \longrightarrow & \mathcal{G}(H) & \longrightarrow & H \\
& \| & & \downarrow & & \downarrow{\scriptstyle \rho} \\
1 \longrightarrow & \mathbb{G}_m & \longrightarrow & \mathrm{Aut}^{\mathbb{G}_m}(L) & \longrightarrow & \mathrm{Aut}(Y).
\end{array}
$$

The closed points of $\mathcal{G}(H)$ are the pairs (h, ϕ) such that $h \in H$ and $\phi \in \mathrm{Aut}^{\mathbb{G}_m}(L)$ lifts h. This is the same as the pairs (h, ψ) such that $h \in H$ and $\psi : L \to h^*(L)$ is an isomorphism of line bundles. In particular, the homomorphism $\mathcal{G}(H) \to H$ is surjective if and only if L is *H-invariant* in the sense that $h^*(L) \cong L$ for all $h \in H$; then $\mathcal{G}(H)$ is a central extension of H by $\mathbb{G}_m$.

The Theta groups constructed by Mumford can easily be checked to match this construction. Following [Mum08], let Y be an abelian variety, L a line bundle on Y, and $K(L)$ the scheme-theoretic kernel of the polarization homomorphism

$$\varphi_L : Y \to \hat{Y}, \quad y \mapsto \tau_y^*(L) \otimes L^{-1}.$$

Then $K(L)$ acts on Y by translations, and this action leaves L invariant. The corresponding central extension

$$1 \to \mathbb{G}_m \to \mathcal{G}(L) \to K(L) \to 1$$

where $\mathcal{G}(L) := \mathcal{G}(K(L))$, is just Mumford's Theta group.

6.4 Homogeneous vector bundles over an abelian variety

From now on we fix A to be an abelian variety; and by a vector bundle we mean a vector bundle over A, unless otherwise stated.

We say that the vector bundle $p : E \to A$ is *homogeneous* if, for each $a \in A$, $\tau_a^*(E) \cong E$ as vector bundles over A. This is equivalent to the condition that the image of the induced map

$$p_* : \text{Aut}^{\mathbb{G}_m}(E) \to \text{Aut}(A)$$

contains all translations τ_a, $a \in A$; these form the neutral component of $\text{Aut}(A)$ by Proposition 4.3.2. Since $\text{Aut}^{\mathbb{G}_m}(E)$ is a group scheme, locally of finite type, the above condition is in turn equivalent to the surjectivity of the restriction

$$\text{Aut}^{\mathbb{G}_m}(E)^o_{\text{red}} \to \text{Aut}^o(A) = A.$$

Now recall that E has a largest anti-affine group of automorphisms,

$$\mathcal{G} := \text{Aut}_{\text{ant}}(E),$$

which is contained in $\text{Aut}^{\mathbb{G}_m}(E)$ (see Proposition 5.5.4). It follows that $\mathcal{G}$ is the largest anti-affine subgroup of $\text{Aut}^{\mathbb{G}_m}(E)^o_{\text{red}}$. Also, any connected affine subgroup of $\text{Aut}^{\mathbb{G}_m}(E)^o_{\text{red}}$ is mapped to a point under the homomorphism p_* to the abelian variety A. In view of the Rosenlicht decomposition, it follows that the surjectivity of the above homomorphism is equivalent to the surjectivity of

$$p_* : \mathcal{G} \to A.$$

Note that we have the following commuting diagram by the definition of $\mathcal{G}$:

$$
\begin{array}{ccc}
\mathcal{G} & \hookrightarrow & \mathrm{Aut}^{\mathbb{G}_m}(E) \\
\downarrow{\scriptstyle p_*} & & \downarrow \\
A & \hookrightarrow & \mathrm{Aut}(A).
\end{array}
$$

In particular, the action of $\mathcal{G}$ on A by translations arises from an action on E such that the above diagram commutes, which precisely means that E is $\mathcal{G}$-*linearized*.

Thus, a homogeneous vector bundle $p : E \to A$ yields a short exact sequence of commutative group schemes

$$0 \to H \to \mathcal{G} \to A \to 0$$

and a $\mathcal{G}$-linearized vector bundle E over the homogeneous space $\mathcal{G}/H \cong A$, such that $\mathcal{G} = \mathrm{Aut}_{\mathrm{ant}}(E)$. The subgroup scheme H is affine since it sits in the group of bundle automorphisms $\mathrm{Aut}_Y^{\mathbb{G}_m}(E)$. As $\mathcal{G}$ is anti-affine, we get an anti-affine extension of A in the sense of Section 5.5. Moreover, identifying the fibre E_0 over $0 \in A$ with k^n and noticing that H acts by linear automorphisms on the fibre E_0, we get a representation $\rho : H \to \mathrm{GL}_n$. In fact, ρ is uniquely defined up to conjugacy in GL_n (which corresponds to choosing an isomorphism $E_0 \cong k^n$).

Conversely, given such a short exact sequence and representation, we can get a homogeneous bundle by taking the associated bundle over the homogeneous space $\mathcal{G}/H$. Thus we have proved:

Theorem 6.4.1 *There is a bijective correspondence between homogeneous vector bundles over A of rank n and pairs consisting of:*

(i) an anti-affine extension $0 \to H \to \mathcal{G} \to A \to 0$, and

(ii) *a scheme-theoretic faithful representation $\rho : H \to \mathrm{GL}_n$ determined up to conjugacy.*

A vector bundle E of rank n is said to be *unipotent* if one of the following equivalent conditions holds:

(i) E admits a reduction of structure group to the maximal unipotent subgroup of GL_n consisting of upper triangular $n \times n$ matrices with diagonal entries 1.

(ii) E is obtained by iterated extensions of trivial bundles, *i.e.*, it has a filtration by sub-bundles each of the subsequent quotients being trivial.

We may now prove part of Theorem 1.5.1; specifically, the following:

Theorem 6.4.2 *(i) Every homogeneous vector bundle is an iterated extension of line bundles $L_i \in \hat{A}$.*

(ii) Every homogeneous vector bundle has a unique decomposition $E = \oplus_i L_i \otimes E_i$ where L_i's are pairwise distinct line bundles in $\hat{A}$ and E_i's are unipotent vector bundles.

PROOF: (i) By Theorem 6.4.1, we have $E \cong \mathcal{G} \times^H k^n$ for some representation $\rho : H \to \mathrm{GL}_n$. Since H is commutative, by conjugating suitably we may assume that $\rho(H) \subset B_n$, where B_n denotes the group of upper triangular invertible matrices. This immediately gives a filtration of E by sub-bundles, each subsequent quotient $L = (\mathcal{G} \times^H k \to A)$ being given by a character χ of H. We have the following commuting diagram of extensions

$$
\begin{array}{ccccccccc}
0 & \longrightarrow & H & \longrightarrow & \mathcal{G} & \longrightarrow & A & \longrightarrow & 0 \\
 & & \chi \downarrow & & \downarrow & & \mathrm{id} \downarrow & & \\
0 & \longrightarrow & \mathbb{G}_m & \longrightarrow & \mathcal{G}_\chi & \longrightarrow & A & \longrightarrow & 0
\end{array}
$$

which shows that the $\mathbb{G}_m$-bundle associated with L is a group. Thus $L \in \hat{A}$.

(ii) Since H is a commutative affine group scheme, we have $H = D \times U$ where D denotes the diagonalizable part of H, and U the unipotent part (see [DG70, §IV.2.4]). We can decompose the representation k^n of H as a direct sum of weight spaces with respect to D (or eigen spaces), say $k^n = \oplus_i V_i$, each V_i being stable by H. This decomposition naturally gives a decomposition of the associated vector bundle E, say $E = \oplus_i E_i$. Let χ_i be the character (eigen value) of H corresponding to the weight space V_i. Denote the associated line bundle over A corresponding to the character χ_i by L_i (as constructed above). Observe that D acts trivially on $E_i \otimes L_i^{-1}$ thereby giving a representation of the quotient $H/D \cong U$. Since U is unipotent, any such representation is an iterated extension of trivial representations; hence the vector bundle $E_i \otimes L_i^{-1}$ is unipotent. Thus, we have proved (ii). $\qquad\square$

To complete the proof of Theorem 1.5.1, it remains to show that any vector bundle obtained as an iterated extension of algebraically trivial line bundles is homogeneous. This is a consequence of the following:

Proposition 6.4.3 *Any extension of homogeneous vector bundles is homogeneous.*

PROOF: Let $0 \to E_1 \to E \to E_2 \to 0$ be a short-exact sequence of vector bundles such that E_1, E_2 are both homogeneous. By the discussion before Theorem 6.4.1, we have anti-affine groups $\mathcal{G}_1 = \mathrm{Aut}_{\mathrm{ant}}(E_1)$ and $\mathcal{G}_2 = \mathrm{Aut}_{\mathrm{ant}}(E_2)$ such that $\mathcal{G}_i \twoheadrightarrow A$, and E_i is $\mathcal{G}_i$-linearized for $i = 1, 2$. We now construct an anti-affine group $\mathcal{G} \twoheadrightarrow A$ for which the vector bundle E is $\mathcal{G}$-linearized. Set $G := \mathcal{G}_1 \times_A \mathcal{G}_2$. Then G is a commutative group scheme, equipped with a surjective homomorphism to A. Moreover, G acts (via $\mathcal{G}_i$) on E_i for $i = 1, 2$ hence it acts linearly on $\mathrm{Ext}_A^1(E_2, E_1)$, a finite dimensional k-vector space. Let $\mathcal{G}$ be the largest anti-affine subgroup of G (or equivalently of G_{red}^o). Then the representation of $\mathcal{G}$ on $\mathrm{Ext}_A^1(E_1, E_2)$ is trivial by Proposition 5.1.2; in particular, $\mathcal{G}$ fixes the extension giving E. Hence $\mathcal{G}$ acts on E. Moreover, $\mathcal{G} \twoheadrightarrow A$ as can be proved by the argument before Theorem 6.4.1. $\qquad\square$

Our methods also yield a description of unipotent vector bundles in terms of algebraic groups:

Proposition 6.4.4 *Every unipotent vector bundle is homogeneous. Moreover, under the correspondence of Theorem 6.4.1, the unipotent vector bundles correspond to the anti-affine extensions of A by a unipotent group scheme H, and to arbitrary (faithful) representations of H.*

PROOF: The first assertion follows from Proposition 6.4.3. For the second assertion, let E be a non-trivial unipotent vector bundle. Observe that $H^0(A, E \otimes L) = 0$ for any non-zero $L \in \hat{A}$, since $H^0(A, L) = 0$ (as seen in the proof of Proposition 5.3.2) and since E is an iterated extension of trivial line bundles. Let $E = \mathcal{G} \times^H V$ as given by Theorem 6.4.1; write $H = U \times D$ and $V = \oplus V_i$ as in the proof of Theorem 6.4.2. Then as shown in that proof, $E = \oplus L_i \otimes E_i$ where $L_i \in \hat{A}$ are pairwise distinct and E_i are non-zero homogeneous vector bundles associated to representations of U. Since E_i contains a trivial sub-bundle, we have $H^0(A, E_i) \neq 0$, and hence $H^0(A, E \otimes L_i^{-1}) \neq 0$. Thus, L_i is trivial, and $E = E_i$ so that $H = U$ is unipotent. The converse has been noticed in the proof of Theorem 6.4.2. $\square$

Remark 6.4.5 When $\mathrm{char}(k) = 0$, the anti-affine extensions of A by a unipotent group scheme are exactly the quotients of the universal vector extension

$$0 \to H^1(A, \mathcal{O}_A)^* \to E(A) \to A \to 0$$

by subspaces of $H^1(A, \mathcal{O}_A)^*$ (as follows from Proposition 5.4.2).

When $\mathrm{char}(k) > 0$, such extensions correspond to the local subgroup schemes of $\hat{A}$ (as follows from the discussion before Theorem 5.5.3).

6.5 Mukai correspondence

Let A be an abelian variety. By [Muk78, Theorems 4.12, 4.19], there are equivalences of categories as follows:

$$\left\{\begin{array}{c}\text{Homogeneous}\\ \text{vector bundles}\\ \text{over } A\end{array}\right\} \leftrightarrow \left\{\begin{array}{c}\text{Coherent sheaves on}\\ \hat{A} \text{ with finite support}\end{array}\right\}$$

$$\left\{\begin{array}{c}\text{Unipotent}\\ \text{vector bundles}\\ \text{over } A\end{array}\right\} \leftrightarrow \left\{\begin{array}{c}\text{Coherent sheaves on}\\ \hat{A} \text{ supported at } 0\end{array}\right\}$$

To construct these equivalences, one uses the projections from $A \times \hat{A}$ as below:

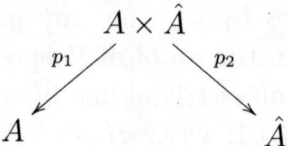

To a coherent sheaf $\mathcal{F}$ on $\hat{A}$, one associates the sheaf

$$(p_1)_*(\mathcal{P} \otimes p_2^*(\mathcal{F}))$$

on $\hat{A}$. Here $\mathcal{P}$ is the the Poincaré line bundle on $A \times \hat{A}$, which is the universal line bundle over $A \times \hat{A}$ (see [Mil86, §9.3]). The first equivalence extends to an equivalence between the bounded derived categories of coherent sheaves on A and $\hat{A}$ (see [Muk81, Theorem 2.2]).

We present an alternative construction of such equivalences, which may be viewed as more elementary, but is non-canonical and only valid in characteristic zero (whereas Mukai's approach is canonical and characteristic-free).

Let $\mathcal{F}$ be a coherent sheaf on $\hat{A}$ with support at 0. Then $\mathcal{F}$ is just a module of finite length over the local ring $\mathcal{O}_{\hat{A},0}$ or equivalently, over its completion $\widehat{\mathcal{O}_{\hat{A},0}}$. The latter is the ring of formal power series $k[[t_1, \dots, t_g]]$, where $t_1, \dots, t_g$ denote local co-ordinates at 0. So $\mathcal{F}$ is a finitely generated module over $k[[t_1, \dots, t_g]]$, killed by some power $(t_1, \dots, t_g)^n$ of the maximal ideal. In other words, $\mathcal{F}$ is a finite-dimensional k-vector space V equipped with g nilpotent endomorphisms $t_1, \dots, t_g$ which commute pairwise. This is in turn equivalent to giving a rational representation

$$\rho : \mathbb{G}_a^g \to \mathrm{GL}(V)$$

of the vector group of dimension g, given by

$$\rho(x_1, \ldots, x_g) = \exp(t_1 x_1 + \cdots + t_g x_g).$$

Further, $t_1, \ldots, t_g$ yield an isomorphism from k^g to the tangent space $T_0(\hat{A})$. Recall that the latter is canonically isomorphic to $\mathrm{H}^1(A, \mathcal{O}_A)$; this yields an isomorphism $\mathbb{G}_a^g \cong \mathrm{H}^1(A, \mathcal{O}_A)^*$.

Clearly, the assignment $\mathcal{F} \mapsto \rho$ induces isomorphisms

$$\mathrm{Hom}_{\hat{A}}(\mathcal{F}_1, \mathcal{F}_2) \cong \mathrm{Hom}(\rho_1, \rho_2)$$

(the space of morphisms of representations). So the category of coherent sheaves on $\hat{A}$ with support at 0 is equivalent to that of finite-dimensional representations of the vector group $\mathrm{H}^1(A, \mathcal{O}_A)^*$. But the latter category is equivalent to that of unipotent vector bundles over A, via the assignment $\rho \mapsto E_\rho := E(A) \times^{\mathrm{H}^1(A,\mathcal{O}_A)^*} V$, where $E(A)$ denotes the universal extension of A (see Section 5.4; recall that $E(A)$ is anti-affine). Indeed, every unipotent vector bundle is isomorphic to a unique E_ρ in view of Proposition 6.4.4 and of Remark 6.4.5. Moreover, for any representations $\rho_i : \mathrm{H}^1(A, \mathcal{O}_A)^* \to \mathrm{GL}(V_i)$, $i = 1, 2$, we have

$$\begin{aligned}
\mathrm{Hom}_A(E_{\rho_1}, E_{\rho_2}) &= \mathrm{H}^0(A, E_{\rho_1}^* \otimes E_{\rho_2}) \\
&= (\mathcal{O}(E(A)) \otimes V_1^* \otimes V_2)^{\mathrm{H}^1(A,\mathcal{O}_A)^*} \\
&= (V_1^* \otimes V_2)^{\mathrm{H}^1(A,\mathcal{O}_A)^*} = \mathrm{Hom}(\rho_1, \rho_2).
\end{aligned}$$

This yields the second equivalence of categories; one can obtain similarly an equivalence between the category of vector bundles of the form $L \otimes U$, where L is a prescribed line bundle in $\hat{A}$ and U a unipotent vector bundle, and the category of coherent sheaves on $\hat{A}$ supported at L. The first equivalence follows from this, in view of Theorem 6.4.2(ii) and of the vanishing of $\mathrm{Hom}_A(L_i \otimes E_i, L_j \otimes E_j)$ for all distinct L_i, L_j in $\hat{A}$ and all unipotent vector bundles E_i, E_j. (This vanishing follows in turn from that of $\mathrm{Hom}_A(L_i, L_j)$, shown for example in the proof of Proposition 5.3.2). $\qquad\square$

Chapter 7

Homogeneous principal bundles over an abelian variety

In this chapter, we obtain structure and classification results for those homogeneous principal bundles over an abelian variety that are homogeneous, and for special classes of such bundles as well. For this, we follow the approach developed in [Bri12].

We begin by extending the classification of homogeneous vector bundles to the setting of principal bundles under a connected affine algebraic group G (Theorem 7.1.3), and we describe their bundle automorphisms (Proposition 7.1.4), thereby proving Theorem 1.6.1.

Then we show that a G-bundle over an abelian variety is homogeneous if and only if so are all associated vector bundles (Theorem 7.2.2). If G is reductive, then it suffices to check the homogeneity of the vector bundle associated to a faithful representation (Corollary 7.2.4). On the other hand, all homogeneous bundles under a connected unipotent group are homogeneous (Corollary 7.2.5).

Next, we obtain a characterization of those homogeneous G-bundles that are indecomposable in the sense of Balaji, Biswas and Nagaraj [BBN05], *i.e.*, that admit no reduction of structure group to a proper Levi subgroup (Proposition 7.3.1); here G is assumed to be reductive, and $\mathrm{char}(k) = 0$. We also introduce notions of simple, resp. irreducible G-bundles, and investigate these notions in the setting of homogeneous bundles. In particular, we show that when G is semi-simple, a G-bundle over an abelian variety is homogeneous and irreducible if and only if its adjoint vector bundle has vanishing cohomology in all degrees (Theorem 7.3.5).

7.1   Structure

As in the previous Chapter, we fix an affine algebraic group G and an abelian variety A. Let $\pi : X \to A$ be a G-bundle. We say that π is *homogeneous* if $\tau_a^*(X) \cong X$ as G-bundles, for any $a \in A$. Like for vector bundles, homogeneity is equivalent to the condition that the image of the induced homomorphism

$$\pi_* : \mathrm{Aut}^G(X) \to \mathrm{Aut}(A)$$

contains the subgroup A of translations.

Example 7.1.1 Let $G = \mathrm{GL}_n$ and $p : E := X \times^G k^n \to A$ be the vector bundle of rank n associated to the G-bundle π. Then π *is homogeneous if and only if so is p* (as follows from the isomorphism $\mathrm{Aut}^G(X) \cong \mathrm{Aut}^{\mathbb{G}_m}(E)$ obtained in Section 6.3.

To study homogeneous G-bundles, one can easily reduce to the case where G is connected (see [Bri12] for details). So we shall assume throughout that G *is connected*; then X is a smooth variety. By an argument as in Section 6.4, one obtains the following:

Proposition 7.1.2 *A G-bundle $\pi : X \to A$ is homogeneous if and only if π_* restricts to a surjective homomorphism $\mathrm{Aut}_{\mathrm{ant}}(X) \to A$.*

The arguments of Section 6.4 also adapt readily to yield:

Theorem 7.1.3 *There is a bijective correspondence between homogeneous G-bundles over A and pairs consisting of the following data:*

(i) an anti-affine extension $0 \to H \to \mathcal{G} \to A \to 0$,

(ii) a scheme-theoretic faithful homomorphism $\rho : H \to G$ determined upto conjugacy.

This correspondence assigns to any pair as above, the associated bundle $\pi : G \times^H \mathcal{G} \to \mathcal{G}/H = A$, where H acts on $\mathcal{G}$ by multiplication, and on G via right multiplication through ρ.

We now describe the group scheme of bundle automorphisms, $\mathrm{Aut}_A^G(X)$, of any homogeneous G-bundle:

Proposition 7.1.4 *With the notations of the above Theorem, denote by $C_G(H)$ the scheme-theoretic centralizer of the image of H in G. Then the action of $C_G(H)$ on G by right multiplication induces an action on $X = G \times^H \mathcal{G}$ by bundle automorphisms. Moreover, the resulting morphism*

$$\varphi : C_G(H) \to \mathrm{Aut}_A^G(X)$$

is an isomorphism of group schemes. The pre-image of A under the homomorphism $\pi_* : \mathrm{Aut}^G(X) \to \mathrm{Aut}(A)$ is isomorphic to the quotient $(C_G(H) \times \mathcal{G})/H$, where H is embedded diagonally in $C_G(H) \times \mathcal{G}$.

PROOF: Let $C_G(H)$ act on $G \times \mathcal{G}$ by $z \cdot (g, \gamma) := (gz^{-1}, \gamma)$. This action commutes with the G-action by left multiplication on itself, and with the H-action by $h \cdot (g, \gamma) := (g\rho(h)^{-1}, h\gamma)$. Thus, the action of $C_G(H)$ descends to an action on $G \times^H \mathcal{G}$ which lifts the trivial action on $A = \mathcal{G}/H$. This proves the first assertion.

For the second assertion, recall that for any scheme S, the group

$$\mathrm{Aut}_A^G(X)(S) = \mathrm{Aut}_{A \times S}^G(X \times S)$$

is isomorphic to $\mathrm{Hom}^G(X \times S, G)$, the group of G-equivariant morphisms $f : X \times S \to G$ where G acts on $X \times S$ via the given action on X and the trivial action on S, and G acts on itself by conjugation. This isomorphism, θ, assigns to any such f the morphism $(x, s) \mapsto (f(x, s) \cdot x, s)$.

Since X is the categorical quotient of $G \times \mathcal{G}$ by H, we obtain

$$\mathrm{Hom}^G(X \times S, G) \cong \mathrm{Hom}^{G \times H}(G \times \mathcal{G} \times S, G)$$

where G acts on $G \times \mathcal{G} \times S$ via left multiplication on itself. Thus,

$$\operatorname{Hom}^G(X \times S, G) \cong \operatorname{Hom}^H(\mathcal{G} \times S, G).$$

Now G being affine, we have

$$\operatorname{Hom}^H(\mathcal{G} \times S, G) \cong \operatorname{Hom}^H(\mathcal{O}(G), \mathcal{O}(\mathcal{G} \times S))$$

where the right-hand side denotes the homomorphisms of H-algebras. But $\mathcal{O}(\mathcal{G} \times S) \cong \mathcal{O}(S)$ since $\mathcal{G}$ is anti-affine; this yields

$$\operatorname{Hom}^H(\mathcal{O}(G), \mathcal{O}(\mathcal{G} \times S)) \cong \operatorname{Hom}^H(\mathcal{O}(G), \mathcal{O}(S)) \cong \operatorname{Hom}^H(S, G).$$

Moreover, $\operatorname{Hom}^H(S, G) \cong \operatorname{Hom}(S, C_G(H))$, since H acts trivially on S. Putting these isomorphisms together, we obtain an isomorphism

$$\operatorname{Hom}^G(X \times S, G) \cong \operatorname{Hom}(S, C_G(H))$$

which assigns to any $\phi : S \to C_G(H)$ the morphism $(g, \gamma, s) \mapsto g\phi(s)g^{-1}$. This isomorphism composed with θ is our morphism φ; hence φ is an isomorphism as required.

To show the final assertion, recall that $\pi_*(\mathcal{G}) = A$, so that $\pi_*^{-1}(A) = \ker(\pi_*)\mathcal{G} = C_G(H)\mathcal{G}$ where the product is taken in $\operatorname{Aut}^G(X)$. Moreover, $C_G(H)\mathcal{G} \cong (C_G(H) \times \mathcal{G})/(C_G(H) \cap \mathcal{G})$, and the scheme-theoretic intersection $C_G(H) \cap \mathcal{G}$ is the kernel of the restriction of π_* to $\mathcal{G}$. But this kernel is just H; this completes the proof. $\square$

Remark 7.1.5 For any G-bundle $\pi : X \to A$, the map π *is the Albanese morphism of* X. Indeed, since G is affine and connected, any morphism (of varieties) from G to an abelian variety B is constant (see [Mil86, Corollary 3.9]), and hence any morphism $f : X \to B$ is G-invariant. Thus, f factors through π, as required. In particular, π has an intrinsic description.

If in addition π is homogeneous, then we obtain another morphism

$$\phi : X = G \times^H \mathcal{G} \to G/\rho(H)$$

which is a $\mathcal{G}$-bundle since H is identified to a subgroup of G via ρ. In fact, ϕ *is the affinization morphism, i.e.,* the canonical morphism $X \to$ Spec $\mathcal{O}(X)$. To see this, note that

$$\mathcal{O}(X) \cong (\mathcal{O}(G) \otimes \mathcal{O}(\mathcal{G}))^H \cong \mathcal{O}(G)^H \cong \mathcal{O}(G/\rho(H)).$$

Here, the first isomorphism holds by the identification of X as $G \times^H \mathcal{G}$, the second follows from anti-affineness of $\mathcal{G}$, and the third is classical. Also, since G is affine and H is commutative, then $G/\rho(H)$ is *quasi-affine*, i.e., the natural morphism $G/H \to$ Spec $\mathcal{O}(G/H)$ is an open immersion (this follows from a result of Grosshans, [Gro97, Theorem 2.1], when H is an algebraic group; the general case of a group scheme is an easy consequence, see [Bri12] for details). In particular, the morphism ϕ is also intrinsic.

Example 7.1.6   *(i) Homogeneous bundle with non-connected group of automorphisms.* Let char$(k) \neq 2$ and let G be the special orthogonal group associated to the quadratic form $x^2 + y^2 + z^2$; then $G = \mathrm{SO}_3 \cong \mathrm{PGL}_2$. Let H be the subgroup of diagonal matrices in G, then $H \cong \mathbb{Z}/2\mathbb{Z} \times \mathbb{Z}/2\mathbb{Z}$. Finally, let E be an elliptic curve. Then the 2-torsion subgroup scheme E_2 is isomorphic to H in view of our assumption on char(k). Thus, we may form the associated bundle $\pi : X := G \times^H E \to E/H$ over the elliptic curve $E/H =: A$. Then X is the homogeneous G-bundle over A associated to the extension $0 \to H \to E \to A \to 0$ and to the inclusion of H into E. Hence by Proposition 7.1.4, the group of bundle automorphisms is isomorphic to $C_G(H)$. Now $C_G(H) = H$ is a non-trivial finite group. In particular, $\mathrm{Aut}_A^G(X)$ is non-connected.

In fact, the full automorphism group scheme, $\mathrm{Aut}^G(X)$, is also non-connected. To see this, it suffices to show that the image of that group scheme under π_* contains $(-1)_A$. But the automorphism id $\times (-1)_\mathcal{G}$ of $G \times \mathcal{G}$ being G-equivariant and H-invariant (since $-h = h$ for all $h \in H$), it descends to a G-automorphism of X which lifts $(-1)_A$.

(ii) *Homogeneous bundle with non-reduced automorphism group scheme:* Let $\operatorname{char}(k) = 2$ and $G = SL_2$. Let H be the subgroup of G generated by the matrix $\left(\begin{smallmatrix} 1 & 1 \\ 0 & 1 \end{smallmatrix}\right)$. Then $H \cong \mathbb{Z}/2\mathbb{Z}$ and $C_G(H) \cong \mu_2 \times \mathbb{G}_a$, where μ_2 (the scheme-theoretic kernel of the square map of $\mathbb{G}_m$) is identified with the scheme-theoretic center of G, and $\mathbb{G}_a$ is viewed as the subgroup of matrices $\left(\begin{smallmatrix} 1 & t \\ 0 & 1 \end{smallmatrix}\right)$. Next, choose an ordinary elliptic curve E, *i.e.*, E has a point of order 2; then this point is unique. This yields an embedding of H into E and, in turn, a homogeneous G-bundle $X := G \times^H E$ over $A := E/H$, as in the above Example. Then $\operatorname{Aut}_A^G(X) = C_G(H)$ by Proposition 7.1.4 again. Hence $\operatorname{Aut}_A^G(X)$ is non-reduced. It follows that $\operatorname{Aut}^G(X)$ is non-reduced as well: indeed, the group scheme

$$\pi_*^{-1}(A) \cong C_G(H) \times^H E \cong \mu_2 \times (\mathbb{G}_a \times^{\mathbb{Z}/2\mathbb{Z}} E)$$

is connected, and hence is the neutral component of $\operatorname{Aut}^G(X)$. Also, $\pi_*^{-1}(A)$ is non-reduced, since so is μ_2.

7.2   Characterizations

The following proposition gives a characterization of homogeneous principal bundles in terms of algebraic groups:

Proposition 7.2.1 *A G-bundle $\pi : X \to A$ is homogeneous if and only if π is trivialized by an anti-affine extension $\mathcal{G} \to A$.*

PROOF: Let X be a homogeneous G-bundle. From Theorem 7.1.3 we know that $X \cong G \times^H \mathcal{G}$ and $A \cong \mathcal{G}/H$, with notations as given there. Let ψ be the composition of the projection $p_2 : G \times \mathcal{G} \to \mathcal{G}$ followed by the quotient morphism $\phi : \mathcal{G} \to A$. Then ψ is invariant under the H-action on $G \times \mathcal{G}$ given by $h \cdot (g, x) := (gh^{-1}, hx)$. So ψ factors through X. This yields a commutative square

$$
\begin{array}{ccc}
G \times \mathcal{G} & \xrightarrow{\;p_2\;} & \mathcal{G} \\
\big\downarrow{\scriptstyle q} & & \big\downarrow{\scriptstyle \phi} \\
X & \xrightarrow{\;\pi\;} & A,
\end{array}
\qquad\qquad (7.1)
$$

where q denotes the quotient morphism by H. Since both the vertical arrows are H-bundles, the square is cartesian. Thus, π is trivialized by the anti-affine extension $\phi : \mathcal{G} \to A$.

To prove the converse, let $\phi : \mathcal{G} \to A$ be an anti-affine extension which trivializes π. Then we have a similar cartesian square as in (7.1). Thus, H acts on $G \times \mathcal{G} \cong X \times_A \mathcal{G}$ by G-equivariant automorphisms, so that p_2 is H-equivariant. So this action is given by $h \cdot (g, \gamma) = (f(g, h, \gamma), h\gamma)$ for some morphism $f : G \times H \times \mathcal{G} \to G$. As G is affine and $\mathcal{G}$ is anti-affine, so $f(g, h, \gamma) = \psi(g, h)$ for some $\psi : G \times H \to G$. The G-equivariance condition yields $\psi(g, h) = g\psi(e_G, h)$. Thus, H acts on G via right multiplication through a homomorphism

$$\rho : H \to G, \quad h \mapsto \psi(e_G, h).$$

Hence $X \cong G \times^H \mathcal{G}$ which is clearly homogeneous. $\qquad \square$

Theorem 7.2.2 *The following conditions are equivalent for a G-bundle $\pi : X \to A$:*

(i) *π is homogeneous.*

(ii) *For any representation $\rho : G \to \mathrm{GL}(V)$, the associated vector bundle $p : E_V = X \times^G V \to A$ is homogeneous.*

(iii) *For any irreducible representation $\rho : G \to \mathrm{GL}(V)$, the associated vector bundle is homogeneous.*

(iv) *For a faithful representation $\rho : G \to \mathrm{GL}(V)$ such that $\mathrm{GL}(V)/\rho(G)$ is a quasi-affine variety, the associated vector bundle is homogeneous. (Such a representation always exists, see Lemma 7.2.3).*

PROOF: (i) $\implies$ (ii): We adapt the argument in Section 6.3. Note that any automorphism $\varphi \in \mathrm{Aut}^G(X)$ gives rise to a G-automorphism $\varphi \times \mathrm{id}$ of $X \times V$. This in turn induces an automorphism φ' of E_V. We therefore have a natural homomorphism

$\Phi : \mathrm{Aut}^G(X) \rightarrow \mathrm{Aut}^{\mathbb{G}_m}(E_V)$ which sends φ to φ'. Moreover, we have a commuting diagram:

$$\begin{array}{ccc} \mathrm{Aut}^G(X) & \xrightarrow{\ \pi_* \ } & \mathrm{Aut}(A) \\ \Big\downarrow{\scriptstyle\Phi} & & \Big\downarrow{\scriptstyle\mathrm{id}} \\ \mathrm{Aut}^{\mathbb{G}_m}(E_V) & \longrightarrow & \mathrm{Aut}(A) \end{array}$$

Since the image of π_* contains A, so does the image of p_*. Thus, p is homogeneous.

(ii) $\Longrightarrow$ (iii) is immediate.

(iii) $\Longrightarrow$ (iv) Every representation of G is an iterated extension of irreducible representations. Hence every associated vector bundle is an iterated extension of vector bundles associated to irreducible representations. Now, any extension of homogeneous bundles is homogeneous (see Proposition 6.4.3). Thus, extending the bundle from irreducible representations gives a homogeneous one.

(iv) $\Longrightarrow$ (i) By Example 7.1.1 (or the above commuting diagram), the $\mathrm{GL}(V)$-bundle $X \times^G \mathrm{GL}(V) \rightarrow A$ is homogeneous. This is equivalent to the surjectivity of the homomorphism

$$\mathcal{G} := \mathrm{Aut}_{\mathrm{ant}}(X \times^G \mathrm{GL}(V)) \rightarrow A$$

in view of Proposition 7.1.2. We now claim that the natural map

$$f : X \times^G \mathrm{GL}(V) \rightarrow \mathrm{GL}(V)/\rho(G)$$

is $\mathcal{G}$-invariant. To see this, observe that the anti-affine group $\mathcal{G}$ acts trivially on the algebra $\mathcal{O}(X \times^G \mathrm{GL}(V)) = (\mathcal{O}(X) \otimes \mathcal{O}(\mathrm{GL}(V)))^G$, by Proposition 5.1.2. Thus, $\mathcal{G}$ acts trivially on the subalgebra $\mathcal{O}(\mathrm{GL}(V)/\rho(G)) = \mathcal{O}(\mathrm{GL}(V))^G$. Since $\mathrm{GL}(V)/\rho(G)$ is quasi-affine, its points are separated by regular functions, which implies the claim.

By the claim, the fibre $f^{-1}(\rho(G)) \simeq X$ over the identity coset is stable under the action of $\mathcal{G}$. This gives a homomorphism $\mathcal{G} \rightarrow \mathrm{Aut}^G(X)$ such that the following diagram commutes,

$$\begin{array}{ccc} \mathcal{G} & \longrightarrow & \mathrm{Aut}^G(X) \\ \Big\downarrow & & \Big\downarrow \\ A & \lhook\joinrel\longrightarrow & \mathrm{Aut}(A). \end{array}$$

This immediately yields the homogeneity of $\pi : X \to A$. $\qquad\square$

Lemma 7.2.3 *Let G be an affine algebraic group. Then there exists a faithful finite-dimensional representation $\rho : G \to \mathrm{GL}(V)$ such that the variety $\mathrm{GL}(V)/\rho(G)$ is quasi-affine.*

PROOF: This is a consequence of a theorem of Chevalley that realizes any homogeneous space under an affine algebraic group as an orbit in the projectivization of a representation (see [Spr09, Theorem 5.5.3]). Specifically, we may assume that $G \subset \mathrm{GL}_n$. By Chevalley's theorem, there exist a representation $\rho : \mathrm{GL}_n \to \mathrm{GL}_N$ and a line $\ell \subset k^N$ such that the isotropy subgroup scheme of ℓ in GL_n is just G. Then G acts on ℓ via a character $\chi : G \to \mathbb{G}_m$. The group $\mathrm{GL}_n \times \mathbb{G}_m$ acts linearly on k^N via $(g, t) \cdot v := t^{-1}\rho(g)v$, and the isotropy subgroup scheme of any non-zero point of ℓ is

$$\{(g, t) \mid g \in G, t = \chi(G)\} \cong G.$$

Thus, for the embedding of G into $\mathrm{GL}_n \times \mathbb{G}_m$ via the homomorphism $g \mapsto (g, \chi(g))$, the variety $(\mathrm{GL}_n \times \mathbb{G}_m)/G$ may be regarded as an orbit in the $\mathrm{GL}_n \times \mathbb{G}_m$-module k^N, and hence is quasi-affine. Next, embed $\mathrm{GL}_n \times \mathbb{G}_m$ into GL_{n+1} via $(g, t) \mapsto \left(\begin{smallmatrix} g & 0 \\ 0 & t \end{smallmatrix}\right)$. Then the variety $\mathrm{GL}_{n+1}/(\mathrm{GL}_n \times \mathbb{G}_m)$ is affine (it may be viewed as the variety of tensors $v \otimes f \in k^{n+1} \otimes (k^{n+1})^*$, where $f(v) = 1$). Moreover, GL_{N+1}/G is a GL_{n+1}-orbit in $\mathrm{GL}_{n+1} \times^{\mathrm{GL}_n \times \mathbb{G}_m} k^N$. The latter variety is the total space of a vector bundle over $\mathrm{GL}_{n+1}/(\mathrm{GL}_n \times \mathbb{G}_m)$, and hence is affine. We conclude that GL_{N+1}/G is quasi-affine. $\quad\square$

Corollary 7.2.4 *Let G be a connected reductive subgroup of GL_n. Then a G-bundle $\pi : X \to A$ is homogeneous if and only if so is the associated vector bundle $X \times^G k^n \to A$.*

PROOF: Since the variety GL_n/G is affine (see for instance, [Ric77]), the assertion follows from the equivalence of statements (i) and (iv) of Theorem 7.2.2. $\qquad\square$

Corollary 7.2.5 *Let $\pi : X \to A$ be a G-bundle, U a connected unipotent normal subgroup of G, and $X \xrightarrow{\phi} Y \xrightarrow{\psi} A$ the factorization of π into a U-bundle followed by a G/U-bundle. Then π is homogeneous if and only if so is ψ.*

In particular, any principal bundle under a connected unipotent group is homogeneous.

PROOF: Note that the homomorphism $\pi_* : \mathrm{Aut}^G(X) \to \mathrm{Aut}(A)$ factors as

$$\mathrm{Aut}^G(X) \xrightarrow{\phi_*} \mathrm{Aut}^{G/U}(Y) \xrightarrow{\psi_*} \mathrm{Aut}(A).$$

Thus, if π is homogeneous, then so is ψ.

To show the converse, we use the equivalence of statements (i) and (iii) of Theorem 7.2.2. Let $\rho : G \to \mathrm{GL}(V)$ be an irreducible representation. Then ρ factors through an irreducible representation $\sigma : G/U \to \mathrm{GL}(V)$, and hence $E_V = X \times^G V$ is isomorphic to $Y \times^{G/U} V$ as vector bundles over A, where G/U acts on V via σ. But the latter vector bundle is homogeneous, since so is ψ. □

To state our next corollary, we recall the notion of characteristic classes in algebraic geometry, after [Vis89]. Given an algebraic group G, a *characteristic class* for G-bundles is a function c that associates to every such bundle $\pi : X \to Y$ a class $c(\pi) \in CH^*(Y)$ (the Chow cohomology ring of X with rational coefficients), such that f is contravariant in the following sense: for any morphism $f : Y' \to Y$, we have $c(\pi') = f^*c(\pi)$, where $\pi' : X \times_Y Y' \to Y'$ denotes the pull-back bundle, and $f^* : CH^*(Y) \to CH^*(Y')$ the pull-back morphism. Characteristic classes form a graded $\mathbb{Q}$-algebra denoted by $\mathcal{C}(G)$; it was shown by Edidin-Graham (see [EG97, EG98]) and Totaro (see [Tot99]) that $\mathcal{C}(G)$ is generated by the Chern classes of associated vector bundles.

Corollary 7.2.6 *All characteristic classes of homogeneous G-bundles over an abelian variety are algebraically trivial.*

PROOF: It suffices to show that all the Chern classes of the vector bundle $p : E_V \to A$, associated to a homogeneous G-bundle

$\pi : X \to A$ and a G-module V, are algebraically trivial. By Theorem 1.5.1, E_V has a filtration with successive quotients being algebraically trivial line bundles L_i. Hence the Chern classes of E_V are the elementary symmetric functions in the classes $c_1(L_i)$, and thus are algebraically trivial. $\qquad\qquad\qquad\qquad\qquad\qquad\qquad\qquad\square$

7.3 Special classes of principal bundles

Throughout this section, we assume that $\mathrm{char}(k) = 0$ and G is a connected reductive algebraic group. Recall that a *Levi subgroup* of G is the centralizer of a subtorus; then the Levi subgroups are exactly the maximal connected reductive subgroups of the parabolic subgroups of G.

For example, when $G = \mathrm{GL}_n$, the Levi subgroups are just the conjugates of the subgroups $\mathrm{GL}_{n_1} \times \cdots \times \mathrm{GL}_{n_r}$, where $n_1 + \cdots + n_r = n$.

Let $\pi : X \to Y$ be a G-bundle over an arbitrary base. Following [BBN05], we say that π is *indecomposable* if it has no reduction of structure group to a proper Levi subgroup of G. For example, when $G = \mathrm{GL}_n$, the indecomposability of π is equivalent to that of the associated vector bundle $p : X \times^G k^n \to Y$.

We now characterize those homogeneous bundles over an abelian variety that are indecomposable, in terms of the data of their classification:

Proposition 7.3.1 *With the notation of Theorem 7.1.3, the following are equivalent for a homogeneous G-bundle $\pi : X \to A$:*

(i) π is indecomposable.

(ii) H is not contained in a proper Levi subgroup of G.

(iii) Every subtorus of $C_G(H)$ is contained in the center Z of G.

(iv) The neutral component $(C_G(H)/Z)^o$ is unipotent.

PROOF: (i) $\implies$ (ii) Let $L \subset G$ be a Levi subgroup. If $H \subset L$, then $X \cong G \times^L Y$, where $Y := L \times^H G$. So π has a reduction of structure group to the L-bundle $\pi \mid_Y : Y \to A$.

(ii) $\implies$ (i) Let L be as above. Assume that π has a reduction of structure group to L and denote by $\varphi : X \to G/L$ the corresponding G-equivariant morphism. Since G/L is affine, φ factors through a G-equivariant morphism $G/H \to G/L$ by Remark 7.1.5. Hence H is contained in some conjugate of L.

(ii)$\Leftrightarrow$(iii) Let T be a subtorus of G, then $H \subset C_G(T)$ if and only if $T \subset C_G(H)$. Moreover, $C_G(T) = G$ if and only if $T \subset Z$. This yields the required equivalence.

Finally, (iii)$\Leftrightarrow$(iv) follows from the characterization of the connected unipotent algebraic groups as those connected affine algebraic groups that contain no non-trivial torus. $\square$

Remark 7.3.2 The equivalence of (i), (iii) and (iv) also follows from a result in [BBN05]: a G-bundle $\pi : X \to Y$ (over an arbitrary base) is indecomposable if and only if $\mathrm{Aut}_Y^G(X)/Z$ contains no non-trivial torus. It is also proved there that every G-bundle has a reduction of structure group to a smallest (up to conjugacy) Levi subgroup L_X; moreover, the corresponding L_X-bundle is indecomposable.

For a homogeneous bundle $\pi : X \to A$, we have $L_X = C_G(T)$ where T is a maximal torus of $C_G(H)$, as can be proved along the lines of the above proof.

Next, we introduce a notion of simplicity for a G-bundle $\pi : X \to Y$, namely, we say that π is *simple* if $\mathrm{Aut}_Y^G(X) = Z$. Here Z is identified to a subgroup scheme of $\mathrm{Aut}_Y^G(X)$ as in Remark 6.1.5. When $G = \mathrm{GL}_n$, this is equivalent to the condition that the associated vector bundle $p : E := X \times^G k^n \to Y$ is simple, *i.e.*, its vector bundle endomorphisms are just scalars.

The structure of simple homogeneous bundles is easily described:

Proposition 7.3.3 *A homogeneous G-bundle $\pi : X \to A$ is simple if and only if G is a torus; then X is a semi-abelian variety over A.*

PROOF: By Proposition 7.1.4, π is simple if and only if we have the equality $C_G(H) = Z$. Since H is commutative, this equality implies that $H \subset Z$, and hence $C_G(H) = G$. Thus, G is commutative, and therefore a torus. So $X = G \times^H \mathcal{G}$ is a connected commutative algebraic group, extension of A by G. Conversely, if X is a semi-abelian variety over A, then $C_G(H) = G = Z$. □

In particular, there are very few simple bundles; we thus introduce a weaker notion. We say that a G-bundle $\pi : X \to Y$ is *irreducible* if the quotient $\mathrm{Aut}_Y^G(X)/Z$ is finite. For instance, the SO_3-bundle constructed in Example 7.1.6(i) is irreducible but not simple. We now obtain group-theoretical characterizations of those homogeneous bundles that are irreducible:

Proposition 7.3.4 *With the notation of Theorem 7.1.3, the following conditions are equivalent for a homogeneous G-bundle $\pi :$ $X \to A$:*

(i) π is irreducible.

(ii) H is diagonalizable and not contained in any proper Levi subgroup of G.

(iii) H is not contained in any proper parabolic subgroup of G.

PROOF: (i) $\Longrightarrow$ (ii) Let U denote the unipotent part of H. Then U is a connected unipotent subgroup of $C_G(H)$, and hence must be trivial. Thus H is diagonalizable. Also, H is not contained in any proper Levi subgroup of G by Proposition 7.3.1.

(ii) $\Longrightarrow$ (iii) Assume that $H \subset P$ for some parabolic subgroup P of G. Then, H being reductive, it is contained in a Levi subgroup of P.

(iii) $\Longrightarrow$ (i) The assumption implies that H is reductive; hence so is $C_G(H)$. Let T be a subtorus of $C_G(H)$; then H is contained in $C_G(T)$, a Levi subgroup of G. Thus, $C_G(T) = G$, *i.e.*, $T \subset Z$. It follows that $C_G(H)/Z$ is a reductive group containing no nontrivial torus, and hence is finite. □

Finally, we characterize those G-bundles $\pi : X \to A$ that are irreducible and homogeneous, in terms of their *adjoint vector bundle* $\mathrm{ad}(\pi)$, *i.e.*, the associated vector bundle to the adjoint representation of G in $\mathrm{Lie}(G)$:

Theorem 7.3.5 *The following conditions are equivalent for a principal bundle $\pi : X \to A$ under a semi-simple algebraic group G:*

(i) π is homogeneous and irreducible.

(ii) $\mathrm{H}^i(A, \mathrm{ad}(\pi)) = 0$ for all $i \geq 0$.

(iii) $\mathrm{H}^0(A, \mathrm{ad}(\pi)) = \mathrm{H}^1(A, \mathrm{ad}(\pi)) = 0$.

PROOF: (i) $\Longrightarrow$ (ii) We have

$$\mathrm{ad}(\pi) \cong X \times^G \mathrm{Lie}(G) = (G \times^H \mathcal{G}) \times^G \mathrm{Lie}(G) \cong \mathcal{G} \times^H \mathrm{Lie}(G).$$

Moreover, H being a finite commutative group, there is an isomorphism of H-modules

$$\mathrm{Lie}(G) \cong \bigoplus_{\lambda \in \hat{H}} \mathrm{Lie}(G)_\lambda,$$

where $\hat{H}$ denotes the character group of H, and $\mathrm{Lie}(G)_\lambda$ the λ-weight space. This yields an isomorphism of vector bundles

$$\mathrm{ad}(\pi) \cong \bigoplus_{\lambda \in \hat{H}} m_\lambda L_\lambda,$$

where L_λ denotes the line bundle over $A = \mathcal{G}/H$ associated to the one-dimensional H-module k_λ with weight λ, and $m_\lambda :=$ $\dim \mathrm{Lie}(G)_\lambda$.

We claim that L_λ is non-trivial for every λ such that $\mathrm{Lie}(G)_\lambda \neq 0$. Indeed, any such λ is non-zero, since $\mathrm{Lie}(G)^H = \mathrm{Lie}(C_G(H)) = 0$. Also, note that

$$\mathrm{H}^0(A, L_\lambda) \cong (\mathcal{O}(G) \otimes k_\lambda)^H \cong \mathcal{O}(G)_{-\lambda}.$$

Since $\mathcal{O}(G)$ is the trivial H-module k, we see that $\mathrm{H}^0(A, L_\lambda) = 0$ and hence L_λ is non-trivial, proving the claim.

By that claim and [Mum08, §III.16], we obtain that $H^i(A, L_\lambda) = 0$ for all $i \geq 0$. This yields the required vanishing.

(ii) $\Longrightarrow$ (iii) is obvious.

(iii) $\Longrightarrow$ (i) The smooth morphism $\pi : X \to A$ yields an exact sequence of tangent sheaves

$$0 \to T_\pi \to T_X \to \pi^*(T_A) \to 0$$

where T_π denotes the relative tangent sheaf. Since π is a G-bundle, T_π is isomorphic to $\mathcal{O}_X \otimes \mathrm{Lie}(G)$ via the action of $\mathrm{Lie}(G)$ on $\mathcal{O}_X$ by vector fields. Moreover, $T_A \cong \mathcal{O}_A \otimes \mathrm{Lie}(A)$, and hence we get an exact sequence

$$0 \to \mathcal{O}_X \otimes \mathrm{Lie}(G) \to T_X \to \mathcal{O}_X \otimes \mathrm{Lie}(A) \to 0.$$

Also, note that all these sheaves are G-linearized (where G acts trivially on $\mathrm{Lie}(A)$), and the morphisms are compatible with the linearizations. So the associated long exact sequence of cohomology begins with an exact sequence of G-modules

$$0 \to \mathcal{O}(X) \otimes \mathrm{Lie}(G) \to H^0(X, T_X) \to \mathcal{O}(X) \otimes \mathrm{Lie}(A)$$
$$\to H^1(X, \mathcal{O}_X) \otimes \mathrm{Lie}(G).$$

Taking G-invariants yields the exact sequence

$$0 \to (\mathcal{O}(X) \otimes \mathrm{Lie}(G))^G \to H^0(X, T_X)^G \to \mathrm{Lie}(A)$$
$$\to (H^1(X, \mathcal{O}_X) \otimes \mathrm{Lie}(G))^G,$$

since G is reductive and $\mathcal{O}(X)^G \cong \mathcal{O}(A) = k$. But we have isomorphisms

$$(H^i(X, \mathcal{O}_X) \otimes \mathrm{Lie}(G))^G \cong H^i(A, \mathrm{ad}(\pi))$$

for all $i \geq 0$, as $\pi : X \to A$ is affine and $(\pi_*(\mathcal{O}_X) \otimes \mathrm{Lie}(G))^G$ is the sheaf of local sections of the adjoint vector bundle $\mathrm{ad}(\pi)$. Since $H^i(A, \mathrm{ad}(\pi)) = 0$ for $i = 0, 1$, it follows that the map

$$\pi_* : H^0(X, T_X)^G \to \mathrm{Lie}(A)$$

is an isomorphism. Thus, the image of $\pi_* : \mathrm{Aut}^G(X) \to \mathrm{Aut}(A)$ contains $\mathrm{Aut}^o(A) = A$, i.e., π is homogeneous. Also, the kernel of

π has dimension 0, *i.e.*, $\mathrm{Aut}_A^G(X)$ is finite; hence π is irreducible. $\square$

Note that the assumption of semi-simplicity cannot be omitted in the above statement. Indeed, for any $\mathbb{G}_m$-bundle $\pi : X \to A$, the adjoint vector bundle is just the trivial line bundle, and hence has non-zero cohomology groups in all degrees $0, \dots, \dim(A)$. But π is homogeneous and irreducible if (and only if) the corresponding line bundle is algebraically trivial.

Bibliography

[Akh95]   D. N. AKHIEZER, *Lie group actions in complex analysis*,
no. E27 in Aspects of Mathematics (Friedr. Vieweg &
Sohn, Braunschweig), 1995.

[Ari60]   S. ARIMA, *Commutative group varieties*, J. Math. Soc.
Japan, 12 (1960), 227–237.

[Ati57a] M. F. ATIYAH, *Complex analytic connections in fibre
bundles*, Tran. Amer. Math. Soc., 85 (1957), 181—209.

[Ati57b] M. F. ATIYAH, *Vector bundles over an elliptic curve*,
Proc. London Math. Soc., 7 (1957), 3, 414–452.

[BB02]   V. BALAJI and I. BISWAS, *On principal bundles with
vanishing Chern classes*, J. Ramanujan Math. Soc., 17
(2002), 3, 187–209.

[BBN05]   V. BALAJI, I. BISWAS and D. NAGARAJ, *Krull-Schmidt
reduction for principal bundles*, J. Reine Angew. Math,
578 (2005), 225–234.

[Bar55]   I. BARSOTTI, *Un teorema di struttura per le varietà grup-
pali*, Atti Accad. Naz. Lincei. Rend. Cl. Sci. Fis. Mat.
Nat., 18 (1955), 43–50.

[BT10]   I. BISWAS and G. TRAUTMANN, *A criterion for homogeneous principal bundles*, Internat. J. Math., 21 (2010),
12, 1633–1638.

[Bla56]   A. BLANCHARD, *Sur les variétés analytiques complexes*,
Ann. Sci. Ecole Norm. Sup., 73 (1956), 3, 157–202.

[BR62]   A. BOREL and R. REMMERT, *Über kompakte homo-
           gene Kählersche Mannigfaltigkeiten*, Math. Ann., 145
           (1961/62), 429–439.

[BoTi71]   A. BOREL and J. TITS, *Éléments unipotents et sous-
           groupes paraboliques de groupes réductifs. I*, Invent.
           math., 12 (1971), 95–104.

[BLR90]   S. BOSCH, W. LÜTKEBOHMERT and M. RAYNAUD,
           Néron models, Ergebnisse der Mathematik und ihrer
           Grenzgebiete (3), vol. 21 (Springer-Verlag, Berlin), 1990.

[Bri09]   M. BRION, *Anti-affine algebraic groups*, J. Algebra, 321
           (2009), 3, 934–952.

[Bri10a]   M. BRION, *On automorphism groups of fiber bundles*,
           arXiv: http://arxiv.org/abs/1012.4606, (2010).

[Bri10b]   M. BRION, *Some basic results on actions of nonaffine
           algebraic groups*, in *Symmetry and spaces*, Progr. Math.,
           vol. 278 (Birkhäuser Boston Inc., Boston, MA), 2010, 1–
           20.

[Bri11]   M. BRION, *On the geometry of algebraic groups and ho-
           mogeneous spaces*, J. Algebra, 329 (2011), 52–71.

[Bri12]   M. BRION, *Homogeneous bundles over abelian varieties*,
           J. Ramanujam Math. Soc., 27 (2012), 1, 107–134.

[Con02]   B. CONRAD, *A modern proof of Chevalley's theorem on
           algebraic groups*, J. Ramanujan Math. Soc., 17 (2002), 1,
           1–18.

[Dem77]   M. DEMAZURE, *Automorphismes et déformations des
           variétés de Borel*, Invent. Math., 39 (1977), no. 2, 179–
           186.

[DG70]   M. DEMAZURE and P. GABRIEL, *Groupes algébriques.
           Tome I: Géométrie algébrique, généralités, groupes com-
           mutatifs* , avec un appendice 'Corps de classes local'

par Michiel Hazewinkel, (Masson & Cie, Éditeur, Paris), 1970.

[EG97] D. EDIDIN and W. GRAHAM, *Characteristic classes in the Chow ring*, J. Algebraic Geom., 6 (1997), 3, 431–443.

[EG98] D. EDIDIN and W. GRAHAM, *Equivariant intersection theory*, Invent. Math., 131 (1998), 3, 595–634.

[Fer03] D. FERRAND, *Conducteur, descente et pincement*, Bull. Soc. Math. France, 131 (2003), 553–585.

[Gri91] R. L. GRIESS, *Elementary abelian p-subgroups of algebraic groups*, Geom. Dedicata, 39 (1991), 3, 253–305.

[Gro97] F. D. GROSSHANS, *Algebraic homogeneous spaces and invariant theory*, Lecture Notes in Mathematics, vol. 1673 (Springer-Verlag, Berlin), 1997.

[Gro60] A. GROTHENDIECK, *Techniques de descente et théorèmes d'existence en géométrie algébrique. I, Généralités. Descente par morphismes fidèlement plats, Séminaire Bourbaki*, vol. 5, 1960, Exposé No. 190.

[HL93] W. HABOUSH and N. LAURITZEN, *Varieties of unseparated flags*, in *Linear algebraic groups and their representations (Los Angeles, CA, 1992)*, Contemp. Math., vol. 153 (Amer. Math. Soc., Providence, RI), 1993, 35–57.

[Har77] R. HARTSHORNE, *Algebraic geometry*, Graduate Texts in Mathematics, vol. 52 (Springer-Verlag, New York), 1977.

[Kam79] T. KAMBAYASHI, *Automorphism group of a polynomial ring and algebraic group action on an affine space*, J. Algebra, 60 (1979), 439–451.

[Mar71] M. MARUYAMA, *On automorphism groups of ruled surfaces*, J. Math. Kyoto Univ., 11 (1971), 89–112.

[Mat63]   H. MATSUMURA, *On algebraic groups of birational trans-formations*, Atti Accad. Naz. Lincei Rend. Cl. Sci. Fis. Mat. Natur., 34 (1963), 151–155.

[MO67]   H. MATSUMURA and F. OORT, *Representability of group functors, and automorphisms of algebraic schemes*, Invent. Math., 4 (1967), 1–25.

[Mat59]   Y. MATSUSHIMA, *Fibrés holomorphes sur un tore complexe*, Nagoya Math. J., 14 (1959), 1–24.

[MM74]   B. MAZUR and W. MESSING, *Universal extensions and one-dimensional crystalline cohomology*, Lecture Notes in Mathematics, vol. 370 (Springer-Verlag, New York), 1974.

[Mil86]   J. S. MILNE, *Abelian varieties*, in *Arithmetic geometry (Storrs, Conn., 1984)* (Springer, New York), 1986, 103–150.

[Miy73]   M. MIYANISHI, *Some remarks on algebraic homogeneous vector bundles*, in *Number theory, algebraic geometry and commutative algebra, in honor of Yasuo Akizuki* (Kinokuniya, Tokyo), 1973, 71–93.

[Mor59]   A. MORIMOTO, *Sur la classification des espaces fibrés vectoriels holomorphes sur un tore complexe admettant des connexions holomorphes*, Nagoya Math. J, 15 (1959), 83–154.

[Muk78]   S. MUKAI, *Semi-homogeneous vector bundles on an abelian variety*, J. Math. Kyoto Univ., 18 (1978), 2, 239–272.

[Muk81]   S. MUKAI, *Duality between $D(X)$ and $D(\hat{X})$ with its application to Picard sheaves*, Nagoya Math. J., 81 (1981), 153-175.

[Mum66]   D. MUMFORD, *On the equations defining abelian varieties. I*, Invent. Math., 1 (1966), 287–354.

[Mum08] D. MUMFORD, *Abelian varieties, Tata Institute of Fundamental Research Studies in Mathematics*, vol. 5 (Published for the Tata Institute of Fundamental Research, Bombay), 2008, with appendices by C. P. Ramanujam and Yuri Manin, Corrected reprint of the second (1974) edition.

[MFK94] D. MUMFORD, J. FOGARTY and F. KIRWAN, *Geometric invariant theory, Ergebnisse der Mathematik und ihrer Grenzgebiete (2)*, vol. 34 (Springer-Verlag, Berlin), third edn., 1994.

[NP11] B. C. NGÔ and P. POLO, *Une démonstration d'un théorème de Chevalley, Rosenlicht et Barsotti*, preprint, (2011).

[Nor83] M. V. NORI, *The fundamental group-scheme of an abelian variety*, Math. Ann., 263 (1983), 3, 263–266.

[Ray70] M. RAYNAUD, *Faisceaux amples sur les schémas en groupes et les espaces homogènes*, Lecture Notes in Mathematics, vol. 119 (Springer-Verlag, New York), 1970.

[Ric77] R. W. RICHARDSON, *Affine coset spaces of reductive algebraic groups*, Bull. London Math. Soc., 9 (1977), 1, 38–41.

[Ros56] M. ROSENLICHT, *Some basic theorems on algebraic groups*, Amer. J. Math., 78 (1956), 401–443.

[Ros58] M. ROSENLICHT, *Extensions of vector groups by abelian varieties*, Amer. J. Math., 80 (1958), 685–714.

[Ros61] M. ROSENLICHT, *Toroidal algebraic groups*, Proc. Amer. Math. Soc., 12 (1961), 984–988.

[Sal03] C. SANCHO DE SALAS, *Complete homogeneous varieties: structure and classification*, Trans. Amer. Math. Soc., 355 (2003), 9, 3651–3667.

[SS09]   C. Sancho de Salas and F. Sancho de Salas, *Principal bundles, quasi-abelian varieties and structure of algebraic groups*, J. Algebra, 322 (2009), 2751–2772.

[Ser88]   J.-P. Serre, *Algebraic groups and class fields*, Graduate Texts in Mathematics, vol. 117 (Springer-Verlag, New York), 1988, translated from the French.

[Ser01]   J.-P. Serre, *Espaces fibrés algébriques*, in *Documents Mathématiques*, vol. 1 (Soc. Math. France, Paris), 2001, 107–139.

[SGA3]   M. Demazure and A. Grothendieck, *Schémas en groupes, (Propriétés générales des schémas en groupes), Séminaire de Géométrie Algébrique du Bois Marie, (1962-1964)*, Documents Mathématiques, vol. 7 (Soc. Math. France, Paris), 2011.

[Spr09]   T. A. Springer, *Linear algebraic groups*, Modern Birkhäuser Classics (Birkhäuser Boston Inc., Boston, MA), second edn., 2009.

[Ste75]   R. Steinberg, *Torsion in reductive groups*, Advances in Math., 15 (1975), 63–92.

[Sum74]   H. Sumihiro, *Equivariant completions*, J. Math. Kyoto Univ., 14 (1974), 1–28.

[Tot99]   B. Totaro, *The Chow ring of a classifying space*, in *Algebraic K-Theory*, Proc. Sympos. Pure Math., vol. 69 (Amer. Math. Soc.), 1999, 249–281.

[Tot11]   B. Totaro, *Pseudo-abelian varieties*, arXiv: http://arxiv.org/abs/1104.0856, (2011).

[Vis89]   A. Vistoli, *Characteristic classes of principal bundles in algebraic intersection theory*, Duke Math. J., 58 (1989), 2, 299–315.

[Wei55]   A. Weil, *On algebraic groups of transformations*, Amer. J. Math., 77 (1955), 355–391.

[Wen93] C. WENZEL, *Classification of all parabolic subgroup-schemes of a reductive linear algebraic group over an algebraically closed field*, Trans. Amer. Math. Soc., 337 (1993), 1, 211–218.

Index

LEARNING DISABILITIES

A non-specialist introduction for
nursing, health and social care